色素増感太陽電池研究者のための

色素データ集
Data book on Dye-sensitized Solar Cells

堀内　保，藤沢潤一，内田　聡　著

シーエムシー出版

はじめに

　地球温暖化対策の観点から，自然エネルギーの利用拡大は待ったなしの状況にある。
　そのような中で，色素増感太陽電池は低コストな次世代の発電デバイスとして各方面から注目を浴びている。本太陽電池のみが持つカラフルな意匠性，透明性，フレキシビリティーといった多くの特色をうまく活かして応用展開していけば，単なる発電素子という立場から，やがては未来のエネルギーを支える産業へと規模が拡大していくのも遠い先のことではないと思われる。実用化し，産業として成り立っていくためには，更なる高効率化・耐久性向上が求められている。そのためには常に新しい材料開発が必要であり，とりわけ色素が果たす役割は大きい。

　本書は，これまでに発表されている色素増感太陽電池用色素67種類について，分子構造や光吸収特性に関わる基礎的物性をまとめたものである。
　データ収集とHOMO/LUMOエネルギー計算・軌道の可視化に際しては藤沢氏の全面的な協力を得た。
　また，単に数値を並べるだけではなく，各種色素開発の変遷については，この分野の最前線に立つ堀内氏より総論を解説頂いた。初期の歴史的な取り組みから始まり，近年性能向上が著しい有機色素の開発に至るまでを余すとこなく，一挙に書ききって頂いた。
　今後さらにデータベースを充実させ，色素増感太陽電池研究の発展に貢献が果たせれば幸いである。

　最後に本書の出版に際して，データ等をご提供頂いた皆様にこの場を借りて感謝致します。

2008年7月末

著者を代表して　内田　聡

目　次

【総論編】

色素増感太陽電池の研究動向

1　はじめに……………………………… 3
2　増感色素について…………………… 3
3　キサンテン系色素…………………… 4
4　ペリレン型増感色素………………… 5
5　銀塩写真用増感色素………………… 6
6　クマリン型色素……………………… 10
7　ポリエン色素………………………… 12
8　ポルフィリン型色素………………… 15
9　フタロシアニン型色素……………… 15
10　ポリマー型色素……………………… 18
11　インドリン型色素…………………… 19
12　チアジアゾール……………………… 20
13　Ar 部位の検討 ……………………… 20
14　インドリン型色素の長波長化……… 22
15　D149, D150 の特性 ………………… 23
16　各色素のサイクリックヴォルタンメトリー…………………………………… 24
17　D149 と D102 の CV 測定結果の比較
　　………………………………………… 25
18　各色素の TG-DTA の測定結果 …… 26
19　インドリン型のまとめ……………… 26
20　D-205 について……………………… 27
21　インドリン型色素の応用例………… 28
22　固体化の検討………………………… 28
23　おわりに……………………………… 29

【色素データ編】

Anthocyanine ………………………… 37
N3 ……………………………………… 38
N621 …………………………………… 40
N712 …………………………………… 41
N719 …………………………………… 42
N749 …………………………………… 44
N820 …………………………………… 45
N823 …………………………………… 46
N845 …………………………………… 47
N886 …………………………………… 48
N945 …………………………………… 49
K9 ……………………………………… 50
K19 …………………………………… 51
K23 …………………………………… 52
K51 …………………………………… 53
K60 …………………………………… 54

K66	55	Eosin Y	84
K69	56	Mercurochrome	85
K73	57	MK-2	86
K77	58	D77	87
Z316	60	D102	88
Z907	61	D120	90
Z907Na	63	D131	91
Z910	64	D149	93
WMC217	65	D150	95
WMC234	66	HRS-1	96
WMC236	67	JK-1	97
WMC239	68	JK-2	98
WMC273	69	D190	99
CYC-B1	70	D205	100
Perylene	71	ZnTPP	101
MKX-?	72	H_2TC_1PP	102
NKX-2311	73	H_2TC_4PP	103
NKX-2510	74	Phthalocyanine Dye	104
NKX-2569 3d	75	Phthalocyanine Dye	105
NKX-2586	77	Phthalocyanine Dye (TT-1)	106
NKX-2587	78	Pendant type Polymer	107
NKX-2677	79	Polythiophene Dye (P3TTA)	108
NKX-2697	80	Cyanine Dye (C1-D)	109
NKX-?	81	Cyanine Dye (SQ-3)	110
NKX-2753	82	Cyanine Dye (B1)	111
NKX-2883	83		

総論編

色素増感太陽電池の研究動向

堀内　保[*1], 藤沢潤一[*2], 内田　聡[*3]

1　はじめに

色素増感システムは比較的歴史が古く，1887年 James Moser らが光励起された色素から半導体への電荷移動（色素増感）現象を報告している[1]。

更に1976年，坪村・松村らが多孔性 ZnO 電極とローズベンガル色素の組み合わせによる光電変換の結果を発表した。当時，効率は単色光で2.5%（at 563nm）と低く，また耐久性も十分ではなかったため，photocell という表現が使われた。しかしながら今日見られるような半導体多孔質電極，色素，ヨウ素電解液という色素増感太陽電池を構成する全ての要素を備えていた（図1）[2]。

その後1991年，Grätzel らが多孔性 TiO_2 電極と Ru 金属錯体の組み合わせで効率7.12%を報告（AM 1.5, 100mW・cm^{-2}）した（図2）[3]。これにより初めてこのシステムが太陽電池として認識されるようになった。

2　増感色素について

色素増感太陽電池において，増感色素は最も重要なキーマテリアルの一つである。その増感色素は Ru 錯体を始めとする「金属錯体」と，金属を全く含まない「有機色素」の大きく2種類に分類される。その特徴を以下に要約する。

・金属錯体
　長所…安定性，吸収波長が広い
　短所…レアメタル（材料価格の高騰の可能性），分子設計自由度が低い
・有機色素

[*1]　Tamotsu Horiuchi　㈱リコー　先端技術研究所（tamotsu.horiuchi@nts.ricoh.co.jp）
[*2]　Jun-ichi Fujisawa　東京大学　先端科学技術研究センター（ufujisw@mail.ecc.u-tokyo.ac.jp）
[*3]　Satoshi Uchida　東京大学　先端科学技術研究センター（uchida@rcast.u-tokyo.ac.jp）

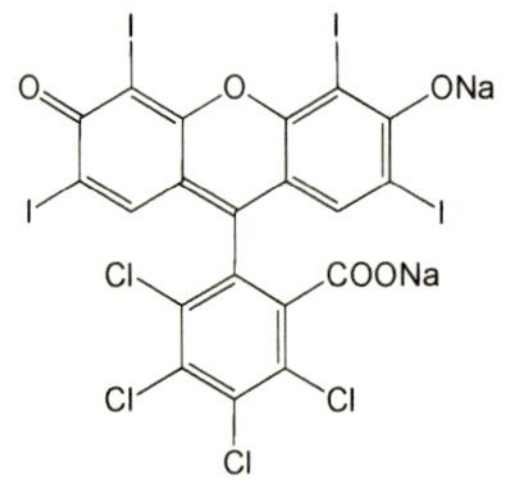

図1 Chemical Structure of Rose Bengal
Semiconductor electrode : porous ZnO, Counter Electrode : Pt, Dye : Rose Bengal 1×10^{-5}M, Electrolyte : KI 0.1M + I_2 0.001M, Efficiency : 1%/100mW・cm^{-2}, IPCE 22%/563nm

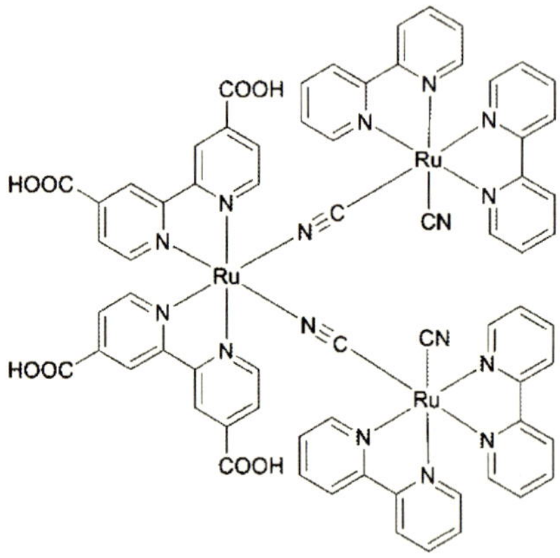

図2 Chemical Structure of Ruthenium complex
Semiconductor electrode : porous TiO_2, Counter Electrode : Pt, Dye : Ru complex 0.5×10^{-4}M, Electrolyte : Tetrapropylammonium Iodide 0.5M + KI 0.02M + I_2 0.04M in EC : AN (80 : 20 vol.%), (V_{OC} 0.681V, J_{SC} 11.46mA・cm^{-2}), FF 0.684, Efficiency : 7.12%/75mW・cm^{-2}, Area 0.5cm^2

長所…モル吸光係数（ε）が高い，鮮やか（吸収波長領域が狭い），低コスト，分子設計自由度が高い

短所…吸収波長域が狭い，開放電圧が低い，耐久性低い

本稿では特に最近性能の進捗が著しい有機色素に焦点を当てて，開発の変遷を概説する。

3 キサンテン系色素

ローズベンガルも含まれる「キサンテン系色素」は非常に安価で入手が容易である。その反面，吸収波長域が狭く耐久性も低い。この骨格の色素は産業総合技術研究所より早くから発表されている。エオシンYを吸着したTiO_2電極では変換効率1.3%（図3）[4]，マーキュロクロームを吸着したZnO電極では変換効率2.5%であった（図4）[5]。

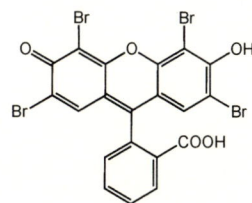

図 3-1 Chemical Structure of Eosin Y
V_{OC} 0.66V, J_{SC} 2.9mA・cm^{-2}, FF 0.67, Efficiency：1.3%

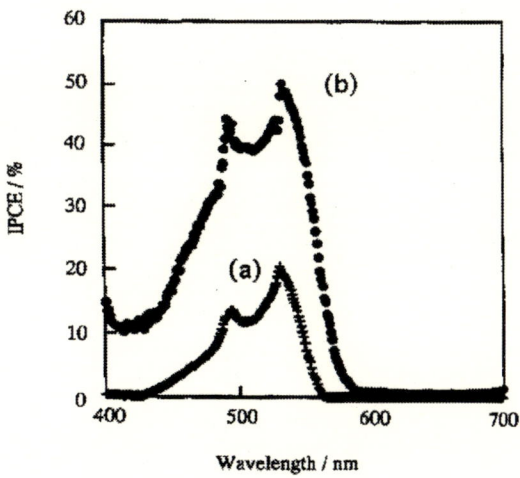

図 3-2 Photocurrent action spectra of EY/TiO$_2$ cell
(a) TiO$_2$ electrode was not treated with TiCl$_4$, (b) TiO$_2$ electrode was treated with TiCl$_4$ aqueous solution for 18h

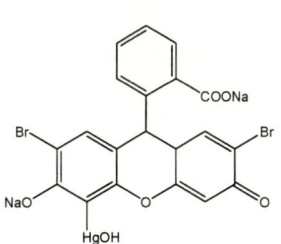

図 4-1 Chemical Structure of Mercurochrome
V_{OC} 0.52V, J_{SC} 7.44mA・cm^{-2}, FF 0.64, Efficiency：2.5%

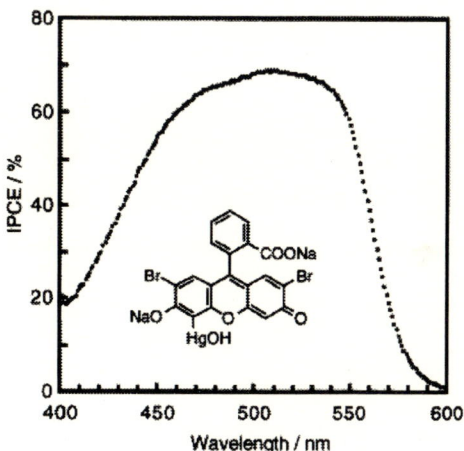

図 4-2 An action spectrum of IPCE for a mercurochromesensitized nanocrystalline ZnO solar cell.

4　ペリレン型増感色素

　赤色顔料として有名なペリレンを用いた検討例としては，NREL より 1997 年に報告されている[6]。Ru 錯体に比較して高いモル吸光係数を持つこと，励起一重項から TiO$_2$ への速い電子移動，の 2 つを期待してペリレンを選択している。電極に多孔質 SnO$_2$，電解液も臭素系を用いている。

ヨウ素溶液を用いているが，電圧が 0.2V しか得られなかったため臭素系電解液を用いたと思われる（図5）。

図5　Chemical Structure of Perylene Dyes
Electrolyte：LiBr 0.5M + Br$_2$ 0.05M + tBP 0.2M in AN
PPCDA on SnO$_2$ Electrode：V_{oc} = 0.45V, J_{sc} = 3.26mA・cm^{-2}, ff = 0.455 Efficiency = 0.89%

5　銀塩写真用増感色素

　光化学の分野において最も研究例が多いのは銀塩写真用の増感色素である。銀塩写真は，目に見える可視光領域に光吸収を有する必要性があり，色素増感型以外の太陽電池でも研究例が報告されている。銀塩写真の増感色素として代表的なものは，図6に示すシアニン色素とメロシアニン色素である。色素増感太陽電池においてメロシアニン色素が登場したのは，富士フイルムから特許が公開されたのが 1999 年[7]，論文発表はノートルダム大とケベック大の共同で発表されたのが同じ 1999 年[8]，産業総合技術研究所と林原生物化学研究所は翌年の 2000 年に報告された[9]。近年では，やはり銀塩写真事業で大手だったコニカミノルタが富士フイルムより出願された特許から外れている骨格を中心に特許出願している。

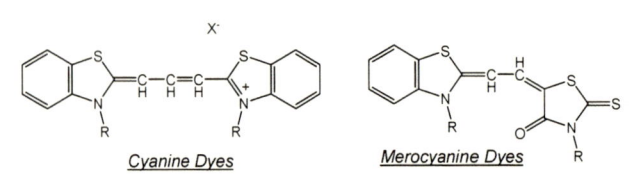

図6　一般的なシアニン色素，メロシアニン色素の骨格

　色素増感型では，メロシアニン色素の方が良好な特性を示す例が多い。産業総合技術研究所と

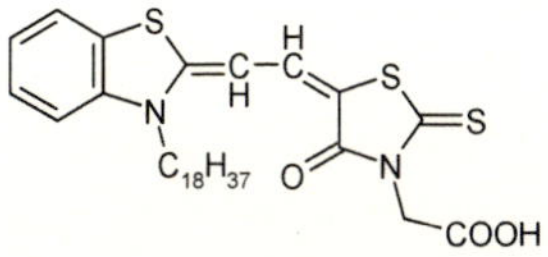

図7 Chemical Structure of Thiazole type Melocyanine Dye

Electrolyte：tetrabutylammonium iodide 0.5M + iodine 0.05M in ethylene carbonate/AN (60/40 v/v). $V_{oc}=0.62V$, $J_{sc}=5.5$ mA·cm^{-2}, ff = 0.54, Efficiency = 2.3%[8]

Electrolyte：LiI 0.1M + 1,2-dimethyl-3-propyl-imidazolium iodide 0.6M + iodine 0.05M + 4-tert-butylpyridine 1M in methoxyacetonitrile. $V_{oc}=0.60V$, $J_{sc}=11.4$ mA·cm^{-2}, ff = 0.65, Efficiency = 4.5%[9]

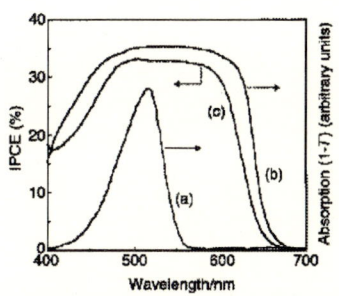

Absorption spectra of Mb(2)-N in ethanol solution (a), Mb(2)-N on TiO$_2$ electrode (b) and IPCE action spectrum of the Mb(2)-N/TiO$_2$ cell (c). The TiO$_2$ electrode and the electrolyte were same as indicated in Table 1.

図8 メロシアニン色素 Mb (2) の UV-Vis 吸収スペクトル

　林原生物化学研究所の報告では，最も有名なチアゾール骨格を有する炭素数18という長鎖アルキル基（ステアリル基）を有するメロシアニン色素で変換効率2.3%，電解液を最適化することで4.2%を達成した[9]。更に2年後には同一色素，同一電解液を用いて4.5%を達成した（図7）[10]。

　チアゾール型メロシアニン色素の特徴は，図8のIPCEスペクトルに見られるように，酸化チタンに吸着すると吸収領域の拡大が生じることである。炭素数18のメロシアニン色素の場合，エタノール中でのλmaxは520nm，吸収端は580nmであるのに対し，TiO$_2$に吸着すると吸収端が670nmにまでレッドシフトする。100nm近いレッドシフトは"J会合体（あるいはJ凝集体）"と呼ばれる特殊な色素間の凝集状態によって形成されていると推測される。J会合体とは，1936年，イーストマン・コダックのJellyが発見したことに因んで，彼の頭文字を取って命名されたものである[11]。その特徴は，①最大吸収スペクトルの長波長化，②スペクトルの線幅が非常に狭くなる，③吸収強度の増大，④ストークスシフトが小さい共鳴蛍光の発現，が挙げられる。ちなみにH会合（H凝集）はHypsochromicの略であり，発見者の名前ではない。このH会合体はJ会合体と逆にブルーシフトすることが知られている。先ほどの吸収波長の大きな広がりは，J会合体だけでなくブルーシフトするH会合体も同時に形成されているためと推測される。

　多孔質膜上に吸着した色素からのストークスシフトの小さい共鳴蛍光の観測は，1996年にGrätzel研より既に報告されている[12]。この論文においては，TiO$_2$上からの発光は観測できていないが，Al$_2$O$_3$上からの発光を観測している。Al$_2$O$_3$上に吸着したメロシアニン色素のλmaxが604nm，発光スペクトルのλmaxが609nmであり，ストークスシフトが非常に小さいことがわかる（図9）。

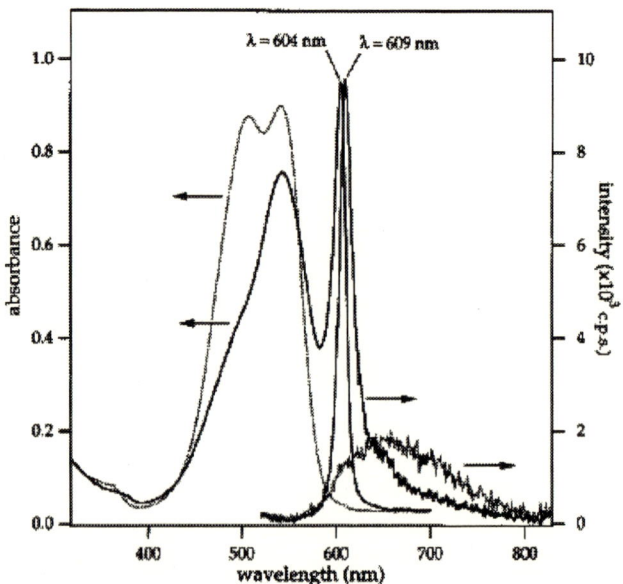

Absorption and emission spectra of Mc2 loaded on a NaOH-impregnated mesoscopic Al₂O₃ film under dry conditions (···) and at a partial water pressure of 19 mbar (—). The excitation wavelength is 490 nm.

図9　メロシアニン色素からの共鳴蛍光スペクトル

シアニン色素も，富士フイルムが早く1999年には特許が公開されている[13]。論文としては，アメリカのChemmotif社より，2000年にカルボン酸を有するシアニン色素が公開されている[14]。最も特性が良好だったのは図10に示す構造のG7と呼ばれるシアニン色素であり，TiO_2電極に吸着した時の吸収スペクトルを見ると，J会合体とH会合体はほぼ等しく形成されているように思われる（図11）。

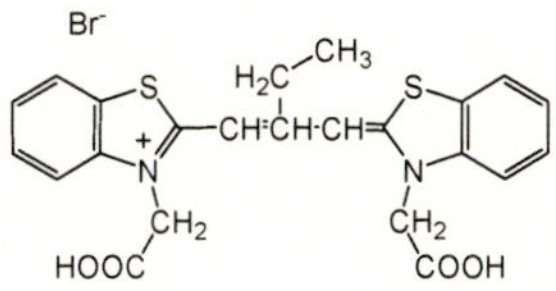

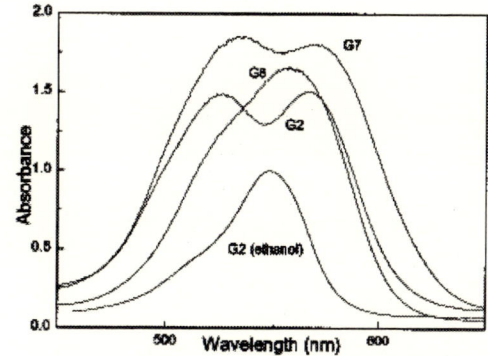

図10 Chemical Structure of Cyanine Dye (G7)

図11 シアニン色素のUV吸収スペクトル

The absorption spectrum of G2 in solution is contrasted with the spectra of G2, G7, and G8 attached onto 4-μm-thick films of nanocrystalline TiO$_2$. The dyes were attached to the solid from a dry ethanol solution.

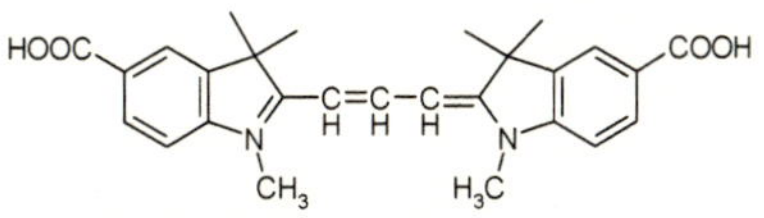

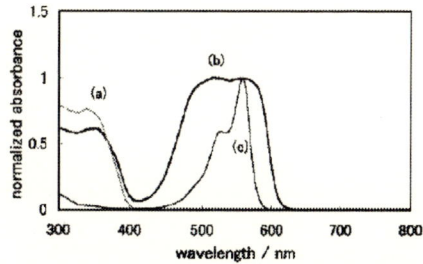

図12 Chemical Structure of Cyanine Dye (C1-D)

Electrolyte : tetrabutylammonium iodide 0.5M +iodine 0.05M in ethylene carbonate/AN (60/40 v/v).
TiO$_2$ electrode : $V_{oc}=0.59V$, $J_{sc}=0.97mA \cdot cm^{-2}$, ff=0.71, Efficiency=0.5% (80mW$\cdot cm^{-2}$)
SnO$_2$ electrode : $V_{oc}=0.39V$, $J_{sc}=2.67mA \cdot cm^{-2}$, ff=0.52, Efficiency=0.7% (80mW$\cdot cm^{-2}$)

図13 スクアリリウム色素のUV-Vis吸収スペクトル

Absorption spectra of (a) TiO$_2$ electrode without dye, (b) C1-D dye on TiO$_2$ electrode and (c) C1-D dye in ethanol solution. The spectra of electrode films were converted from reflection to absorbance by the Kubelka-Munk method.

　その後，産業技術総合研究所と林原生物化学研究所より2001年に報告されたものは[15]，図12の構造のようにカルボン酸の位置が異なるC1-Dと呼ばれるものでのであり，TiO$_2$電極よりもSnO$_2$電極の方が若干ではあるが良好な特性を示すものであった。

　また，中央の連結部位にスクアリリウム酸を用いたものとしては，1999年に中国科学院から報告されたものがある（図13）[16]。変換効率は，図14構造のSQ-3が最も高く2.17%を示しただけでなく，N3色素とSQ-3の混合によって，N3色素単独よりも高い変換効率を示している（N3色素単独：5.87%，混合：6.62%）。

　2007年には，先のシアニン色素を改良した構造のものがGrätzel研より報告されている[17]。インドレニン環のベンゼン環をナフタレン環にすることでより長波長化を狙い，IR領域のみを発電

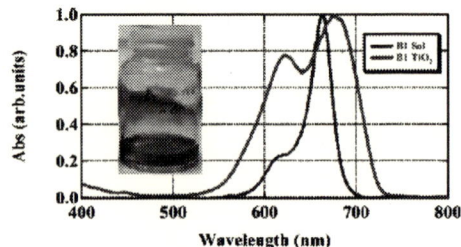

図14 Chemical Structure of Cyanine Dye (SQ-3)
$V_{oc} = 0.54V$, $J_{sc} = 4.4 mA \cdot cm^{-2}$, ff = 0.567, Efficiency = 2.17%

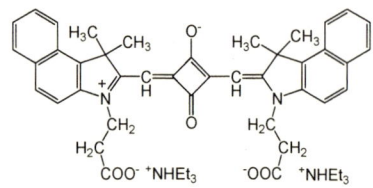

図15 Chemical Structure of Cyanine Dye (B1)
$V_{oc} = 0.59V$, $J_{sc} = 8.6 mA \cdot cm^{-2}$, ff = 0.73, Efficiency = 3.7%

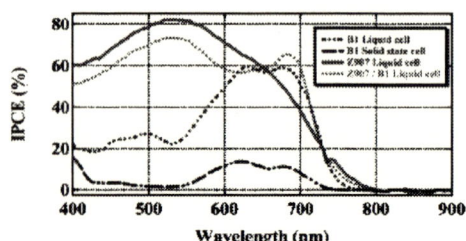

UV-vis curves (normalised) of B1 (blue) in ethanol (6.28 × 10^{-5} M) and adsorbed on an 8 μm TiO$_2$ electrode (red).

IPCEs of three liquid cells sensitised with Z907 (red), B1 (blue) and a Z907/B1 (green) mix, respectively. The IPCE of the solid-state device is also shown (purple line).

図16 スクアリリウム色素のUV-Vis 吸収スペクトル（左）とIPCE スペクトル（右）

する「Photovoltaic Window」を目指しているものである。その結果，溶液中でのλmaxが670〜680nmであるにも係わらず，変換効率3.7%という非常に高い値が得られている（図15）。残念なことは，色素が青色を呈しているため，可視光が透明な「Photovoltaic Window」にはまだ遠いことである（図16）。

6 クマリン型色素

次に，有機色素では特に有名なクマリン型色素を紹介する。

図17に示すクマリン343をTiO$_2$電極上に吸着し，光電流を測定したのもGrätzel研が最初である[18]。この論文では，1SUN（AM1.5, 100mW・cm^{-2}）での変換効率を測定していないものの，IPCEで83.5%（at 440nm）を1989年に報告している。

このクマリン343の改良に成功したのは，産業技術総合研究所と林原生物化学研究所である。2001年には，カルボン酸の部分にメチレン鎖を導入し，末端にシアノ酢酸を導入したNKX-2311が発表された（図18）[19]。このNKX-2311の特徴は，窒素原子と共に環を形成している部位にメチル基を導入していること，さらにクマリン環とカルボン酸と間にメチレン鎖を導入することに

よってπ共役を伸ばし，吸収波長域を広げたところにある。また，TiO_2 電極と吸着するためのカルボン酸として，今では一般的になっているシアノ酢酸を最初に取り入れたのもこの色素である。

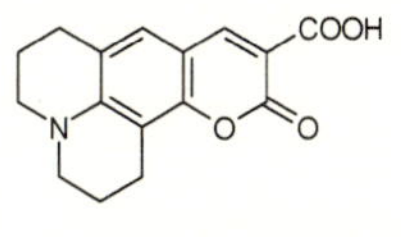

図17 Chemical Structure of Coumarine343
Electrolyte：$HClO_4$ 0.001M + NaI 0.1M in Aqueous

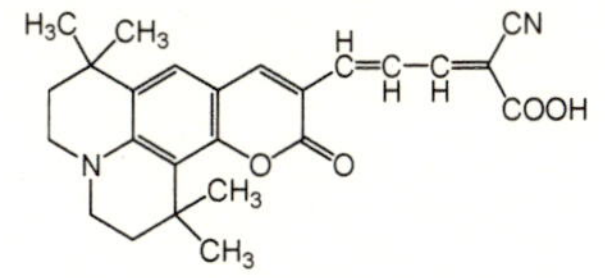

図18 Chemical Structure of Coumarine Dye（NKX-2311）
V_{oc} = 0.63V, J_{sc} = 13.8mA·cm^{-2}, ff = 0.63, Efficiency = 5.6%

この NKX-2311 は LUMO が低いため，電解液に入れる 4-t-ブチルピリジン（4-tBP）との相性が悪く，4-tBP を加えると特性が低下してしまうという弱点があった。

次に，このメチレン鎖をチオフェン環2つに置き換えたものが図19に示す NKX-2677 である。吸収波長域を維持しながら LUMO の低下を防ぐことができたため，電解液に 4-tBP を加えることが可能となり変換効率 7.7% を達成した[20]。NKX-2677 は，電解液の最適化と評価方法（Y44（UV カットフィルター）を通して光照射を60分行い，反射防止フィルムを表面に取り付けることで光電変換特性を評価）を変更することで変換効率 8.3% を達成した[21]。この60分もの光照射後の特性向上のメカニズムについては不明らしいが，熱によるものと推測される（図20）。

更に，2007年には図21に示す構造の NKX-2700 において 8.2% の変換効率を達成した[22]。この色素においても電解液の最適化が行われており，tBP を 0.5M から 0.7M に変更することで効率が向上している（図22）。

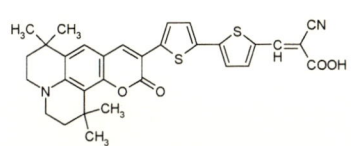

図19 Chemical Structure of Coumarine Dye（NKX-2677）
V_{oc} = 0.73V, J_{sc} = 14.3mA·cm^{-2}, ff = 0.74, Efficiency = 7.7%

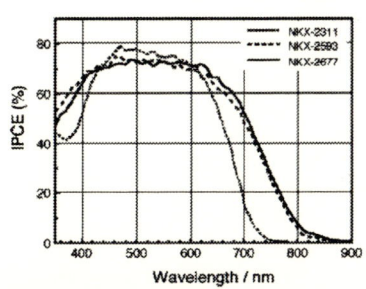

Spectra of monochromatic incident photon-to-current conversion efficiencies (IPCEs) for DSSCs based on coumarin dyes: (…) NKX-2311, (---) NKX-2593, (—) NKX-2677. The electrolyte was the mixture 0.6 M DMPImI–0.1 M LiI–0.05 M I_2 in methoxyacetonitrile.

図20 クマリン色素の IPCE スペクトル

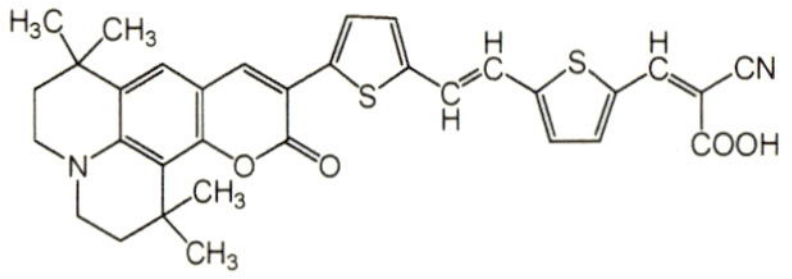

図21　Chemical Structure of Coumarine Dye（NKX-2700）
V_{oc}=0.69V，J_{sc}=15.9mA・cm^{-2}，ff=0.75，Efficiency=8.2%

IPCE as a function of wavelength of incident monochromatic light for TiO_2 electrodes exposed to 0.3 mM NKX-2700 solution containing 0, 10, 60, or 120 mM DCA in a mixture solvent of AN/tert-butanol with volume ratio of 1/1.

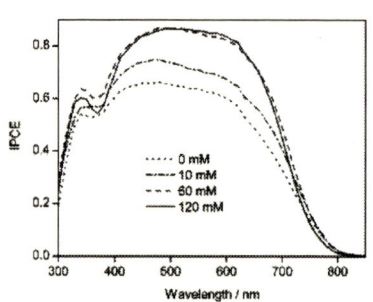

図22　クマリン色素（NKX-2311）の IPCE スペクトル

NKX-2569

NKX-2553

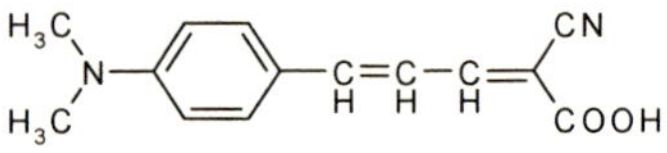

図23　Chemical Structure of Polyene Dye（NKX-2553 and NKX-2569）
NKX-2569：V_{oc}=0.71V，J_{sc}=12.9mA・cm^{-2}，ff=0.74，Efficiency=6.8%

7　ポリエン色素

Ru 錯体の代表である N3 色素は，骨格内に NCS 基からなるドナー部位とビピリジン配位子からなるアクセプタ部位を有する。このドナー－アクセプター構造をよりシンプルに検討した例が図 23 に示す構造のポリエン色素である[23]。

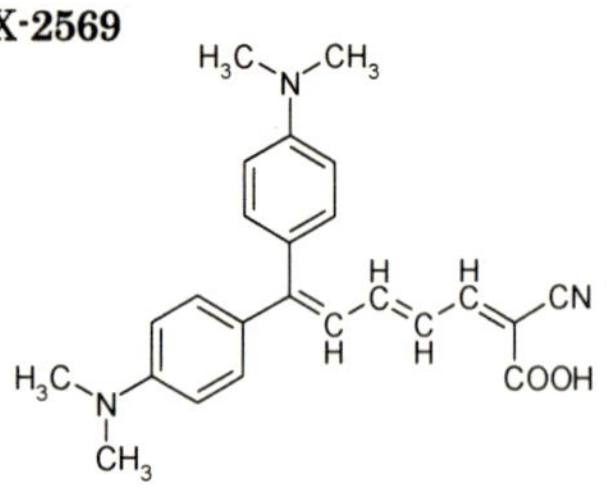

NKX-2553 は非常にシンプルであるが，効率は 5.5%であり，更なる改良が求められた。そこで，よりドナー性が強く，メチレン鎖を延ばした NKX-2569 に改良することによって変換効率6.8%を達成した。

その後，高麗大と Grätzel 研より，図 24 に示す JK-2 が 2006 年に報告された[24]。この色素の改良ポイントは，先のクマリン色素と同様にメチレン鎖をチオフェン環に置き換えているところである。その結果，変換効率 8.0%が得られた。IPCE スペクトルを見ると，吸収端は 700nm であり Ru 錯体やクマリン色素に比較して吸収波長域の更なる改善が必要と思われる（図 25）。

2007 年には韓国の Inha 大学より図 26 の構造の TA-St-CA を用いて 9.1%を達成したと報告

色素増感太陽電池の研究動向

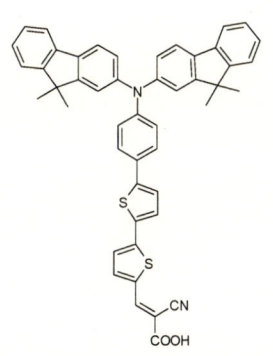

図24 Chemical Structure of JK-2
V_{oc} = 0.75V, J_{sc} = 14.0mA・cm^{-2}, ff = 0.76, Efficiency = 8.0%

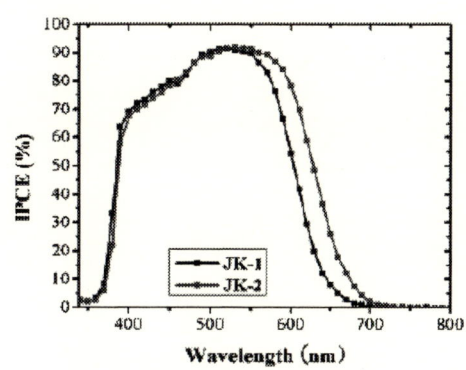

図25 JK色素のIPCEスペクトル

された[25]。IPCEスペクトルを見るかぎり，上記JK-2よりも更に長波長領域の吸収が少なくなっており，今後の改良が期待される（図27）。

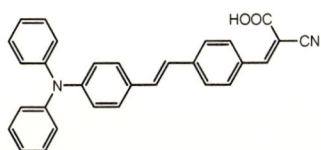

図26 Chemical Structure of TA-St-CA
V_{oc} = 0.743V, J_{sc} = 18.1mA・cm^{-2}, ff = 0.675, Efficiency = 9.1%
Electrolyte：0.7M N-methyl-N-butyl, imidazolium iodide, 0.2M LiI, 50mM iodine, 0.5M tert-butylpyridine in acetonitrile/3-methoxypropionitrile（50：50 V/V）

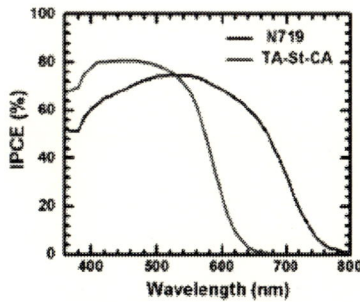

IPCE spectrum for a DSSC based on TA-St-CA.
図27 TA-St-CAのIPCEスペクトル

　また，メチレン鎖をチオフェン環に変更したもう一つの例として，産業技術総合研究所と信州大鈴木研より，カルバゾールにチオフェン環を結合した色素が報告された（図28，29）[26]。この色素は，ドナーとアクセプターの連結基であるチオフェンに，アルキル基を導入したところがポイントである。このアルキル基は，相互作用の強いチオフェンの凝集を防ぐ効果があるため，コール酸のような凝集解離剤を併用する必要がない。2006年の報告では，TiO$_2$膜厚を30μmにすることで変換効率7.7%を，2008年の報告ではMK-2のカルボン酸をトリエチルアンモニウム塩に，更に電解液のヨウ素濃度を0.05Mから0.2Mにすることで8.3%を達成している[27]。

13

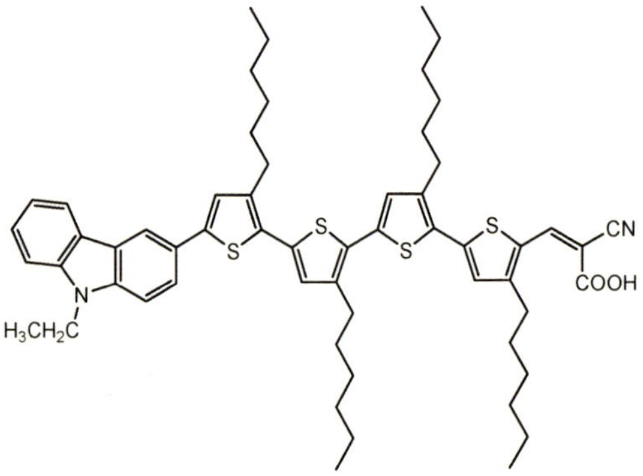

色素増感太陽電池研究者のための色素データ集

図 28 Chemical Structure of MK-2
Electrolyte：0.6M DMPImI + 0.1M LiI + 0.05M I_2 + 0.5M TBP in acetonitrile.
V_{oc} = 0.74V, J_{sc} = 14.0mA·cm^{-2}, ff = 0.74, Efficiency = 7.7%（TiO$_2$ film thickness 30μm）

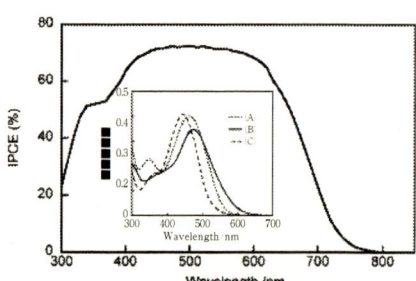

図 29 MK-2 の IPCE スペクトル

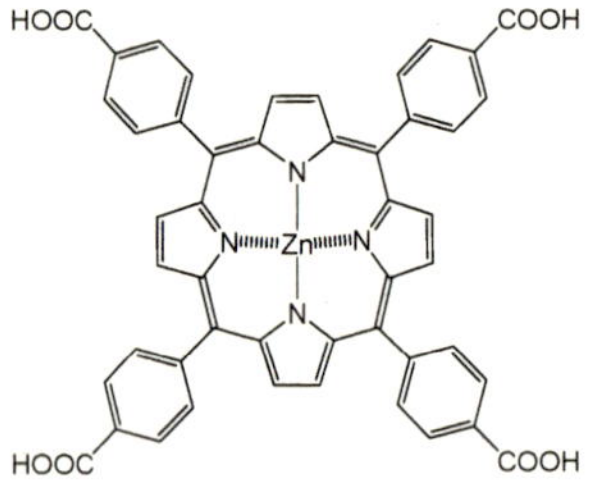

Action spectrum of monochromatic incident photon-to-current conversion efficiency (IPCE) for a DSSC based on MK-2 dye. Thickness of the TiO$_2$ electrode was 30 μm. The inset is absorption spectra of dyes (A) MK-1, (B) MK-2, and (C) MK-3 in chloroform.

図 30 Chemical Structure of ZnTPP

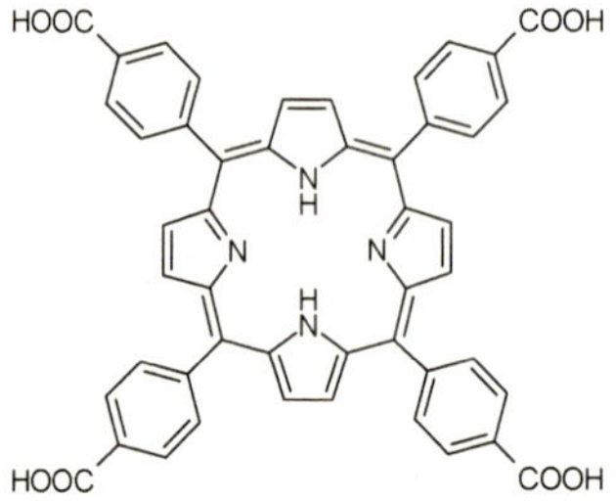

図 31 Chemical Structure of H$_2$TC1PP
V_{oc} = 0.46V, J_{sc} = 3.9mA·cm^{-2}, ff = 0.5, Efficiency = 0.92%

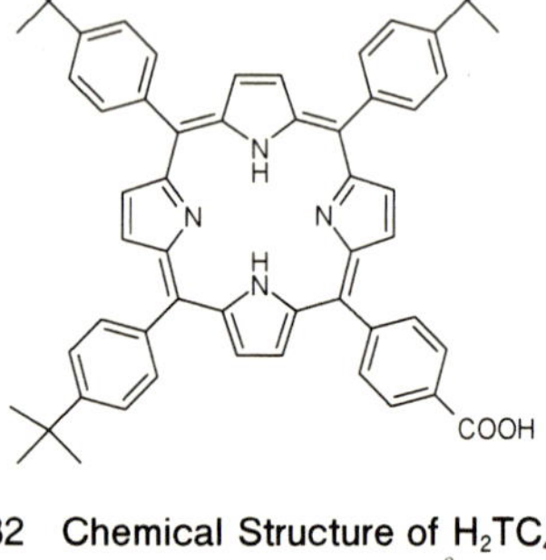

図 32 Chemical Structure of H$_2$TC$_4$PP
V_{oc} = 0.54V, J_{sc} = 5.8mA·cm^{-2}, ff = 0.5, Efficiency = 1.60%

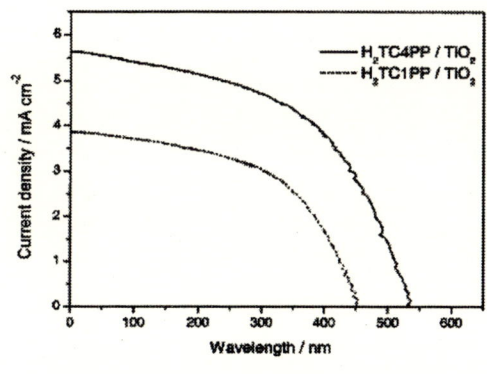

Current density-voltage curves of the porphyrin-sensitized TiO₂ cells.

図33 ポルフィリン色素のIVカーブ

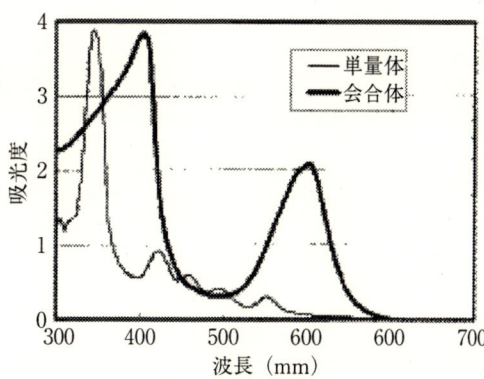

図34 ポルフィリン単量体と会合体のUV吸収スペクトル

8 ポルフィリン型色素

ポルフィリン誘導体の検討は，1987年にGrätzel研より報告されている[28]。ZnTPPCを用い，水溶性電解液を用いた時のアクションスペクトルが，450nmで約50%であった（図30）。後に，クロロフィルを1993年に報告している[29]。

近年では，カルボン酸を4つ有するポルフィリン（H_2TC_1PP，図31）を，t-ブチル基3つに変更したポルフィリン（H_2TC_4PP，図32）が産業技術総合研究所より2002年に報告されている[30]。変換効率から見て，まだまだ改良の余地はあるが，置換基の変更のみで0.92%から1.6%に変換効率が改良されており，中心金属の変更等による更なる特性向上が期待される。t-ブチル基の導入による効果を，色素間の凝集を防ぐ効果があると報告している。

また，東京大学瀬川らにより，ポルフィリンのJ会合体をTiO_2電極の増感色素として検討したことを2006年に報告している[31]。図34にTiO_2電極上にポルフィリンの単量体と会合体を吸着させた時のUV吸収スペクトルを示す。図34から明らかなように，会合体は700nm付近に非常に大きな吸収を有することがわかり，IR領域の光電変換に寄与することが期待される。

9 フタロシアニン型色素

フタロシアニン化合物は非常に堅牢性が高い色素として知られており，古くから有機太陽電池に検討されている。色素増感型においても，堅牢性を生かした耐久性向上の手段として考えている人が多いように思われる。しかしながら，フタロシアニンが高い堅牢性を有しているのは無置

換のものであり，置換基を導入して溶解性を付与したフタロシアニン化合物は，一般的に堅牢性が低下してしまう。

　フタロシアニンをTiO$_2$電極に取り付けた例として，最も早く報告しているのはテキサス大学Bardらによるものであるが[32, 33]，多孔質TiO$_2$上に，溶解性のフタロシアニンを吸着して光電特性を評価したものとしては，1995年，中国東南大学と中国科学院のグループからである[34]。検討したフタロシアニンは図35に示すものであったが，電解液に硫酸ナトリウムとヒドロキノンを使用していた。また，1997年には，同グループよりポルフィリンとフタロシアニンの共吸着が既に報告されていることは驚かされる[35]。

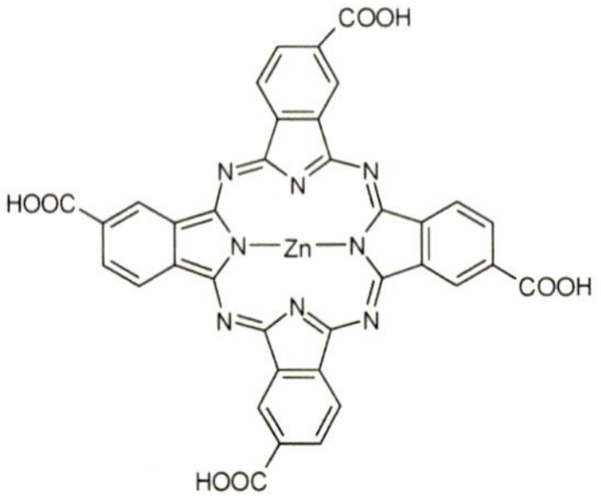

図35　Chemical Structure of Phthalocyanine Dye

　Grätzel研からもフタロシアニンに関する論文は1999年に報告されている[36]。やはり，堅牢性を考慮してフタロシアニンを選択している。1999年の報告では変換効率1.0%，2007年には，アイシンコスモス，インド工科大学との共同で変換効率3%を達成した[37]。改良ポイントは，産業総合研究所におけるポルフィリンと同様でt-ブチル基を3つ，吸着サイトを一つにすることであった。この論文においては，t-ブチル基は溶解性の向上と電荷の再結合抑制効果があると言及している（図36, 37）。

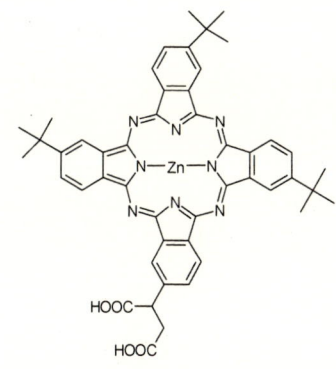

図36 Chemical Structure of Phthalocyanine Dye
$V_{oc} = 0.635V$, $J_{sc} = 6.5mA \cdot cm^{-2}$, ff = 0.743, Efficiency = 3.05%

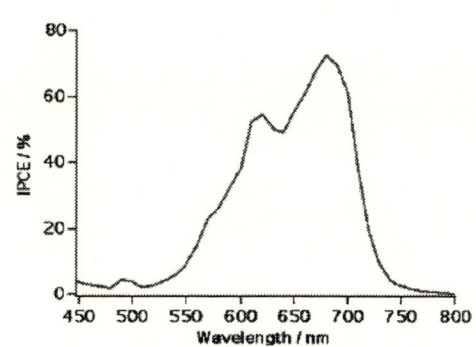

図37 フタロシアニン色素のIPCEスペクトル

　この光電変換特性を見て驚くのは開放電圧（V_{oc}）である。通常，700nm付近にλmaxを有する色素で，ここまで高いV_{oc}を出すことは困難である。また，吸着サイトのカルボン酸とフタロシアニン環との共役が途切れていることも，この色素の特徴である。

　2008年にはマドリッド州立大学とGrätzel研の共同で図38に示す構造のフタロシアニンTT-1を用い，凝集解離剤としてケノデオキシコール酸10mM，更にTiO$_2$膜厚を最適化することで，変換効率3.5%を達成している（図38, 39）[38]。

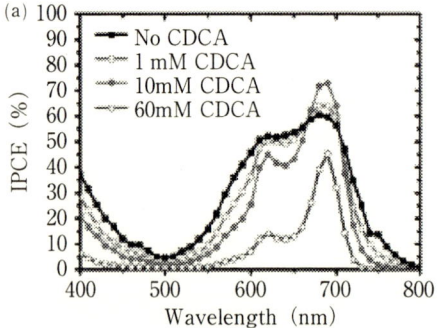

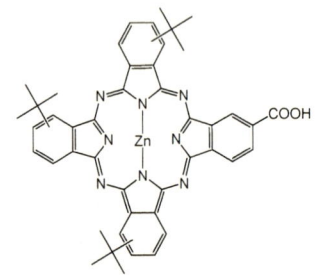

図38 Chemical Structure of Phthalocyanine Dye (TT-1)
$V_{oc} = 0.611V$, $J_{sc} = 7.78mA・cm^{-2}$, ff = 0.751, Efficiency = 3.56%

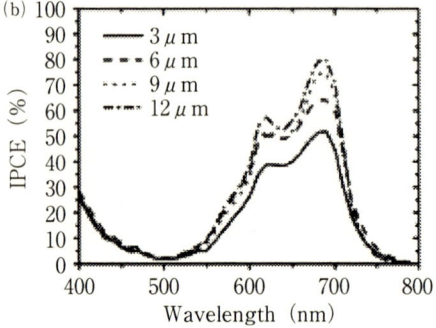

IPCE obtained with a 6μm nanocrystalline TiO$_2$ film supported on ITO conducting glass and derivatized with TT1 as a function of CDCA concentration (a) and thickness of TiO$_2$ film with 1 mM CDCA in dye solution (b). A sandwich type of cell configuration was used to measure this spectrum.

図39 TT-1のIPCEスペクトル

10 ポリマー型色素

変わったアプローチとして，ポリマーを増感色素に応用する検討も報告されている。最初はテキサス大学からの報告でポリイミドにRu錯体をペンダントしたものであった（図40)[39]。

また，導電性ポリマーとしては最もメジャーなポリチオフェンを用いた検討例が阪大柳田研より報告されている[40]。ポリチオフェンは，有機薄膜型太陽電池でも多くの報告例があり，可視光を非常に良好に光電変換する化合物である。色素増感型においても，500nm付近でおよそ60％ものIPCEを有していることは驚かされる。P3TTAは，ヨウ素系電解液にMHImI（1-methyl-3-n-hexylimidazolium iodide）を加えることで効率が向上し，変換効率2.4％を達成した（図41, 42）。

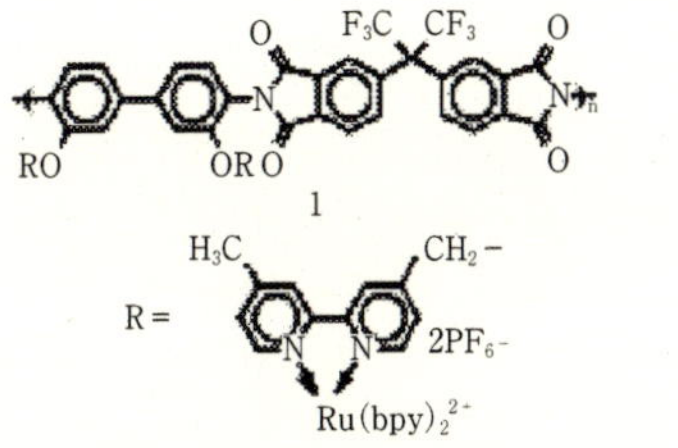

色素増感太陽電池の研究動向

図40 Chemical Structure of Pendant type Polymer
$V_{oc}=0.46V$, $J_{sc}=2.35mA\cdot cm^{-2}$, ff = 0.52, Efficiency = 0.32%

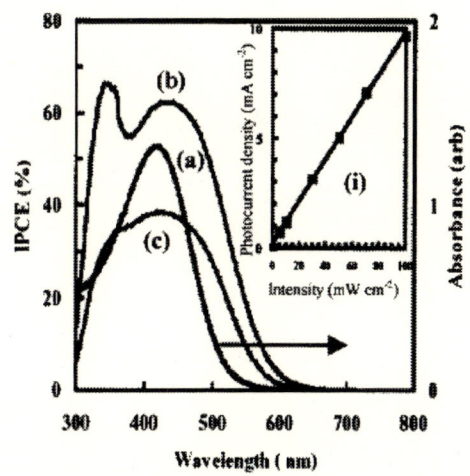

(i) Effect of light intensity on the short-circuit photocurrent of the cell TiO$_2$/P3TAA/electrolyte (a) UV-Vis absorption spectrum of P3TAA in DMSO, (b) IPCE spectrum of TiO$_2$/P3TAA/electrolyte, and (c) IPCE spectrum of SnO$_2$-ZnO/P3TAA/electrolyte.

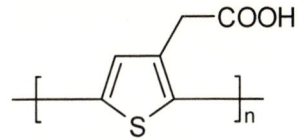

図41 Chemical Structure of Polythiophene Dye (P3TTA)
$V_{oc}=0.40V$, $J_{sc}=9.76mA\cdot cm^{-2}$, ff = 0.61, Efficiency = 2.4%

図42 P3TTA の IPCE スペクトル

11 インドリン型色素

インドリン型色素については，幾つかの学会発表[41]，及び論文に報告されている内容を記載する[42~44]。

三菱製紙も富士フイルムやコニカミノルタと同様に銀塩写真の研究歴があり，増感色素を色素増感型太陽電池へ転用する検討を行った。およそ60種類の増感色素を検討した結果，チアジアゾール環というヘテロ環を有するメロシアニン色素が良好な特性を示すことが分かった。図43は，このチアジアゾール型色素と，一般的によく知られているベンゾチアゾール型色素を比較したものである。実線が溶液中での吸収スペクトル，網線が酸化チタンに吸着した時の吸収スペクトルであり，これらの色素は酸化チタンに吸着することによって，大きくレッドシフトすることが観測された。このレッドシフトも，J会合の形成が示唆されるものである。

チアジアゾール型色素（左）は，ベンゾチアゾール型（右）よりも良好な特性を示した。

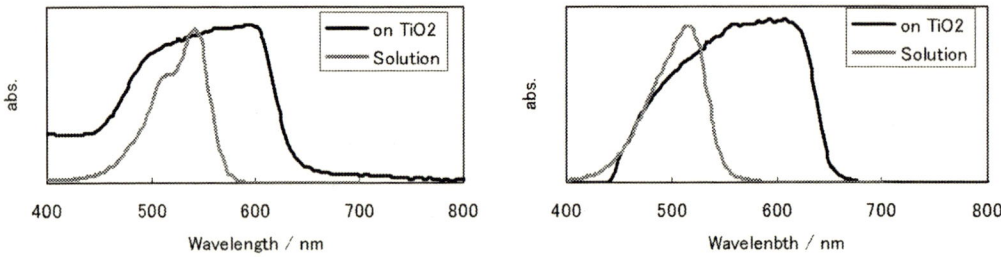

図43 チアジアゾール型色素とベンゾチアゾール型色素の比較
チアジアゾール型：$V_{oc}=0.58V$, $J_{sc}=12.0mA\cdot cm^{-2}$, $ff=0.59$, Efficiency=4.2%
ベンゾチアゾール型：$V_{oc}=0.52V$, $J_{sc}=12.7mA\cdot cm^{-2}$, $ff=0.57$, Efficiency=3.8%

12 チアジアゾール

　チアジアゾールをより詳細に検討した。その結果，R1に置換基を導入した場合，置換基の影響をほとんど受けないことがわかった。そこで，図44のグレーの部分は色素の骨格としては不要ではないかと仮定し，この部分を除去したタイプの色素を検討した。結果，非常に良い特性を示すことが判明した。このような骨格は，写真用増感色素としてはあまり特性が良いものではないことが知られている。写真用増感色素から脱却し，太陽電池用として有望な骨格を発見した瞬間である。

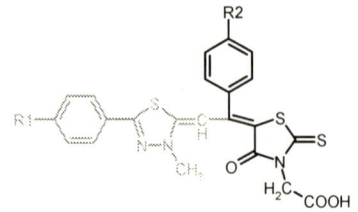

図44 Chemical Structure of Thiaziazole type Merocyanine Dyes

13 Ar部位の検討

　実際には数多くの検討を行っているが，ここでは理解しやすいように3つだけ例を示す。下記Ar部位に，ドナー性の骨格として三級アミンを導入した（図45）。その結果，トリフェニルアミ

ンやカルバゾールに比較して，非常にマイナーなヘテロ環であるインドリンが良い特性を示すことがわかった。図46にIPCEスペクトルを示す。700nm以上の光を吸収できていないことがわかる。この700nm以上の吸収端を長波長化することが高効率化には必須である（図47）。

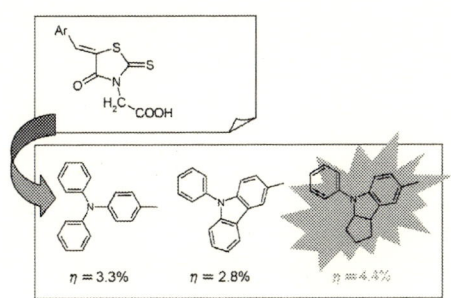

図45 新色素のドナー部位検討例

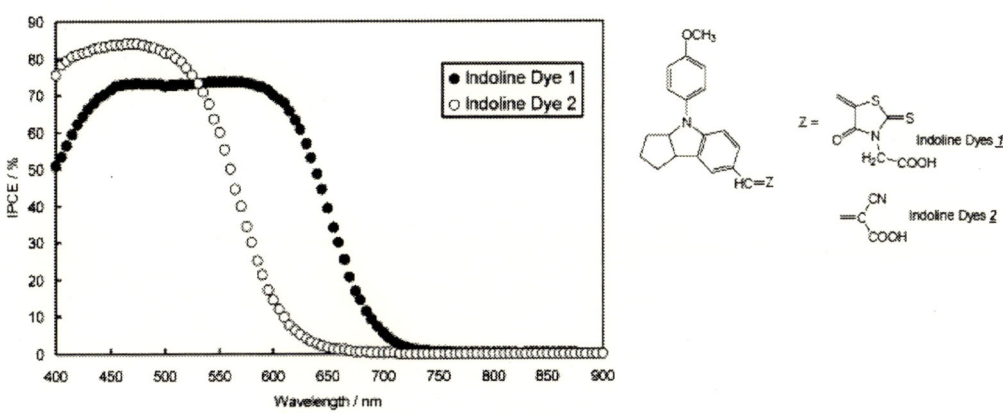

図46 インドリン型色素のIPCEスペクトル

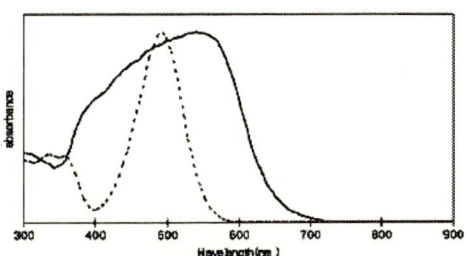

図47 D102のUV-Vis吸収スペクトル

14 インドリン型色素の長波長化

　色素の長波長化は，ドナーとアクセプターの間のメチレン基を増やすことで容易に達成可能であり，それが一般的な検討方法である。しかしながら，長波長化を容易に行える代わり，色素のLUMOが極端に低下してしまい，TiO_2とのレベルのミスマッチが生じ，酸化チタンへ電子を注入することができなくなる可能性が高い。そこで，このドナーとアクセプター間のメチレン鎖を導入することなく，吸収波長域を広げる手法を検討した。

　最初に，インドリン環の結合しているベンゼン環の検討を行った。無置換に比べて，メトキシ基の導入によって効率がわずかではあるが改善されることがわかった。次に，スチリル骨格を導入することによって，π共役の拡大とλmaxの増加，そしてモル吸光係数（ε）が増加することがわかった。近年，新規のRu錯体の検討が世界各地で行われており，非常に多くのRu錯体が報告されている。その大多数は，スチリル基等のπ共役系を導入することによるモル吸光係数（ε）を改良しているが，この検討はその走りと言えるかもしれない。

　次に，アクセプター部位の検討を行った。結論から言うと，アクセプター部位のチオン（＝S）の部分に，新たなもう一つのアクセプターユニットを付け加えることで，レッドシフト化に成功した。この色素は，酸化チタン上に吸着することで紫色に見える。更に，もう一つのアクセプターユニットを導入すると更なるレッドシフトが起こり黒色に見える。逆に，シアノ酢酸を取り付けると吸収波長は短くなり黄色く見える。この中で，開放電圧（V_{oc}）と短絡電流密度（J_{sc}）の関係

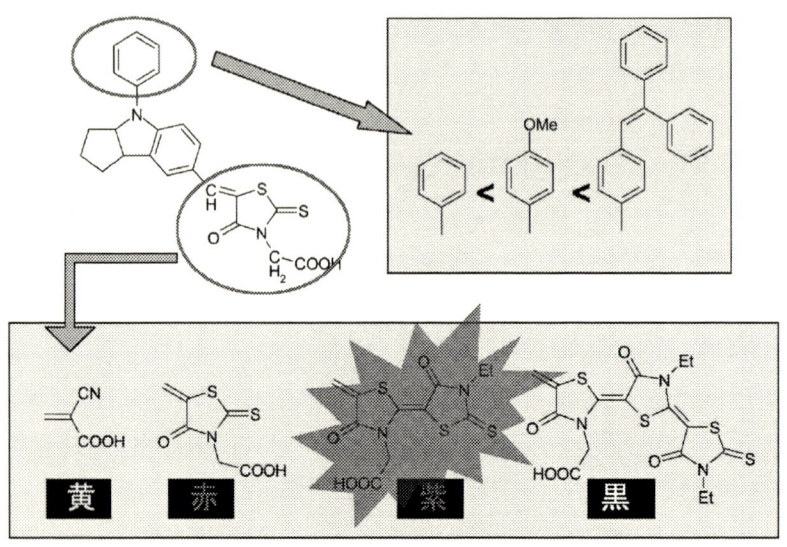

図48　インドリン色素の高効率化検討

から，紫色の色素が最も高い効率を示し，有機色素としては非常に高い変換効率を達成した（図48）。

この骨格の色素が非常に優れているのは，一つのドナーユニットから，様々な吸着サイトを代えるだけで，様々な色調のものを得ることが可能であること，更にその色調のものでは非常に高い効率が得られるという，工業的な実用性が極めて高いことである。例えばクマリン色素やポリエン色素のような骨格においては，吸着サイトをシアノ酢酸からローダニン環に変更すると変換効率が劇的に低下してしまう。

15　D149, D150の特性

図49におけるDye-1がD149, Dye-4がD150である。D149は溶液状態のλmaxが526nm, 酸化チタンに吸着すると541nm。先程のD102に比較してλmaxのシフトが非常に小さい。吸収スペクトルの形状からも，J会合体の存在は少ないものと推測できる。

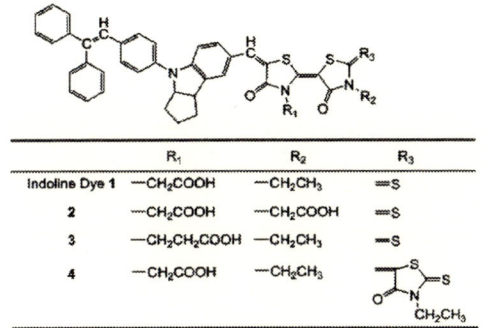
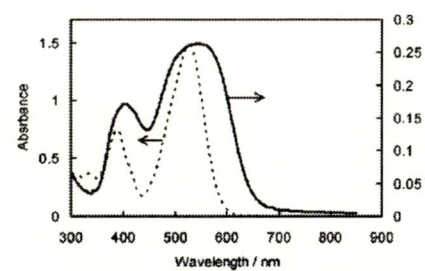

図49　D149, D150の化学構造とUV-Visスペクトル

IPCEスペクトルを観測すると，D149は比較的広い範囲において高いIPCEを有する色素であることが分かる。比較にはD150を載せており，おおよそ60%のIPCEを示している。

D149のIVカーブを図51に示す。一般的に，有機色素はV_{oc}が低いものが多いのだが，D149は凝集解離剤としてコール酸を1mM，電解液に4-tBPを0.05M加えることによって，約700mVの開放電圧が得られている。この高い開放電圧が高い変換効率を得ることに役立っている。

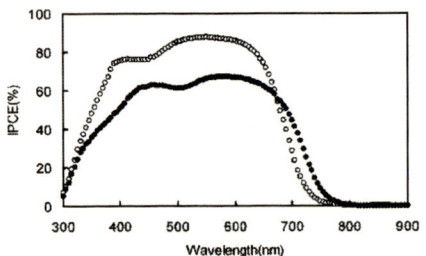
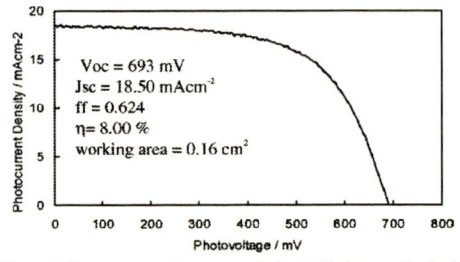

図50 D149 (Dye 1), D150 (Dye 4) のIPCEスペクトル

図51 D149のIVカーブ特性

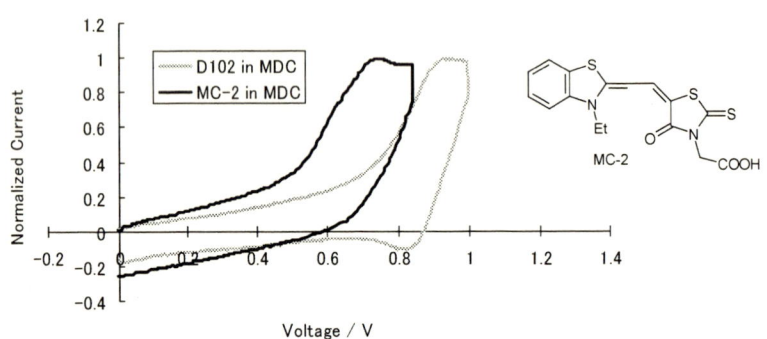

図52 D102とメロシアニン色素 (MC-2) のCV測定図
0.1M Tetra-n-butyl ammonium perchlorate/Methylene Dichloride
D102 in MDC：$E_{1/2}$ = 0.88V (vs. SCE)

16 各色素のサイクリックヴォルタンメトリー

図52に，銀塩写真用増感色素として一般的なチアゾール型のメロシアニン色素（以下MC-2と省略する）とD102の塩化メチレン中でのサイクリックボルタモグラムの測定結果を示す。D102は，酸化された後の還元ピークが現れることが分かる。対してMC-2は還元ピークが現れない。これは，色素の酸化状態が非常に壊れやすいためだと推測している。これに限らず，銀塩写真用の色素は，その用途柄，壊れやすい性質をしている。

また，図53は酸化チタン上にD102を吸着した状態でのCV測定結果である。酸化チタン上では，還元ピークがより明瞭に観測された。同一条件下でMC-2を測定したが，残念ながら酸化電位のピークが発現しなかった。理由は不明で，今後の課題である。

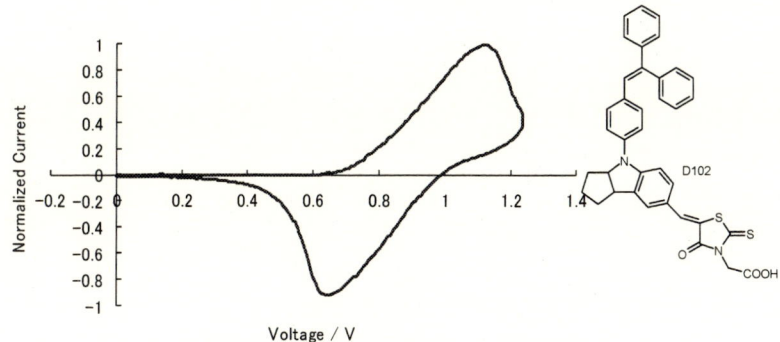

図53　D102（TiO₂電極上）の CV 測定図
0.1M Tetra-n-butyl ammonium perchlolate/Acetnitrile
D102 on TiO₂ : $E_{1/2}$ = 0.89V （vs. SCE）

17　D149 と D102 の CV 測定結果の比較

D149 は，D102 に比較して λmax が長波長シフトしているため，酸化電位も若干低く観測される。しかしながら予想よりも高い値であり，ヨウ素のレベルには適している（図54）。

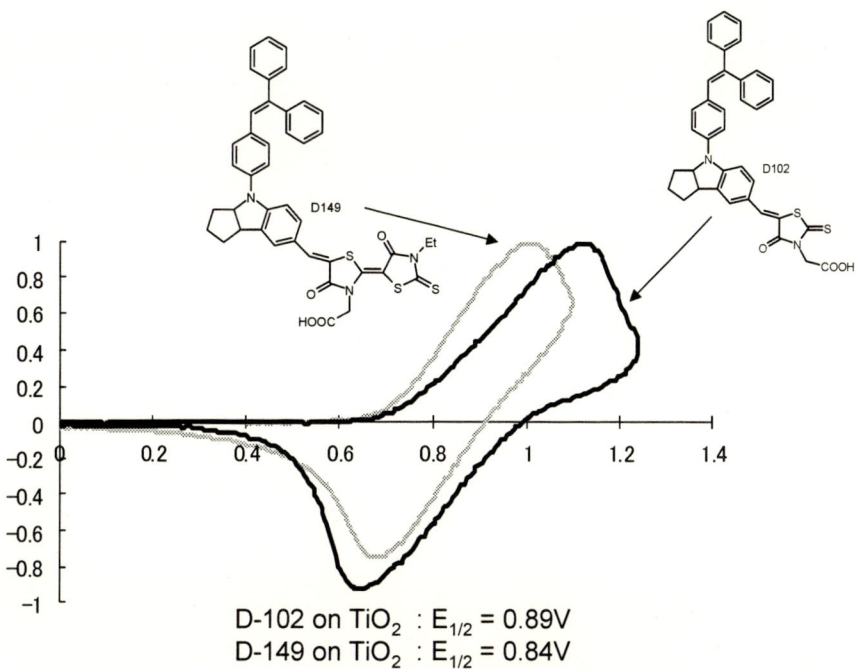

D-102 on TiO₂ : $E_{1/2}$ = 0.89V
D-149 on TiO₂ : $E_{1/2}$ = 0.84V
0.1M Tetra-n-butyl ammonium perchlolate / Acetnitrile

図54　各色素の CV 測定図（TiO₂電極上）

18 各色素のTG-DTAの測定結果

これらの色素の中では，やはりシンプルな構造であるD102が優れた結果を示していることがわかった（図55）。

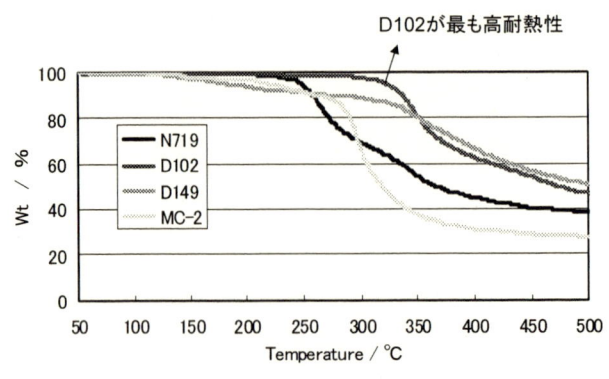

図55　各色素のTG-DTA測定結果

19 インドリン型のまとめ

図56にインドリン色素のまとめを示す。この色素の骨格は，非常に複雑に感じられるが非常に機能性が盛り込まれており無駄がない。Aの部位はπ共役系の広げることによるλmaxのレッドシフト化とモル吸光係数（ε）の増加，更に，立体的な作用によりヨウ素との再結合を防ぐ役割もある。Bの部位は適度なドナー性の付与である。このドナー性とアクセプター部位のバ

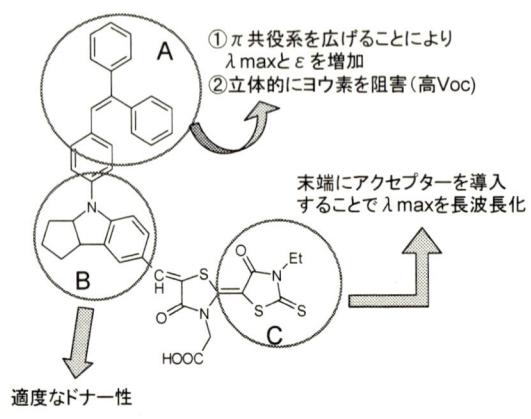

図56　インドリン型色素のまとめ

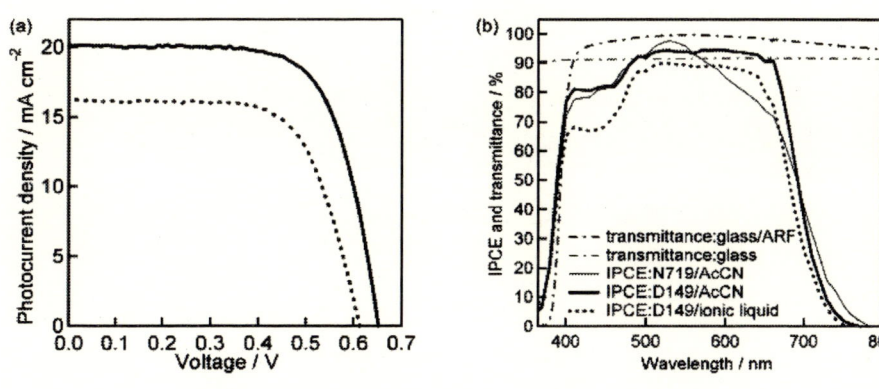

図57　D149のIVカーブ（上）とIPCEスペクトル

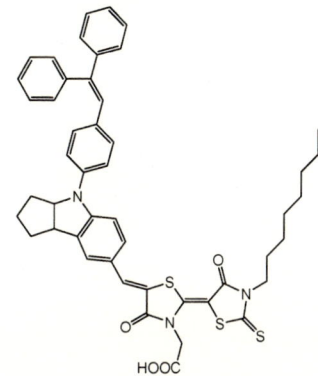

図58　Chemical Structure of D205

ランスは非常に重要である。Cの部位は更にアクセプターを導入することで，λ max をシフトさせる働きがある。

図57は，Grätzel研より報告されたものである[45]。IPCEスペクトルを見てもわかる通り，N719色素とほとんど遜色がない。むしろ，IPCEスペクトル面積全体で考えれば，N719色素を若干上回っていると思われる。

20　D205について

D149の改良として，最近Grätzel研より下記構造のD205が報告された[46]。

D205は，D149にオクチル基を導入したものである（図58）。このオクチル基の効果もあり，イオン液体を用いた時に変換効率で7.2%を達成した。この色素を用い，電解液にヨウ素溶液を

用いたものは 9.5 ％を達成しており，近日中に公開される予定である[47]。

21　インドリン型色素の応用例

　ZnO 電極を用いた場合，Ru 錯体を色素に用いると ZnO を溶解し，電極表面に凝集体を形成することが知られている。インドリン型色素は，このような現象を起さないため ZnO 電極だけでなく，SnO_2 電極でも良好な特性を発揮することが 2003 年に報告されている[48]。SnO_2 電極と D149 の組み合わせに関しては，2005 年に静岡大学より報告されている[49]。N719 と SnO_2 電極の組み合わせが 1.2 ％の変換効率であった時，D149 と SnO_2 電極は 2.8 ％の変換効率を示した。図 59 に示すように，開放電圧の大きさが特に変換効率に寄与している。

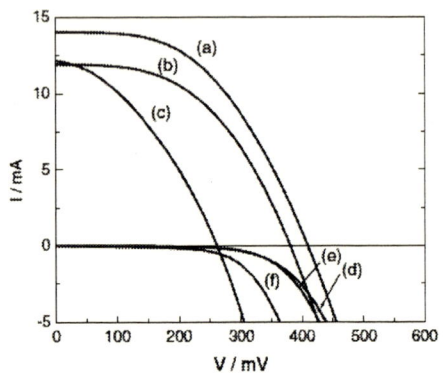

I–V characteristics of SnO_2 cells sensitized with (a) D-149, (b) D-102 and (c) N-719, and dark I–V characteristics of the cells sensitized with (d) D-149, (e) D-102, and (f) N-719.

図 59　各色素の IV カーブ特性（SnO_2 電極）

22　固体化の検討

　メタルフリーの有機色素は，Ru 錯体に比較してモル吸光係数が高いことが特徴の一つである。その高いモル吸光係数を利用すれば，多孔質電極の膜厚を薄くすることも可能となる。その原理を生かし，スピロ型ホール輸送材料（図 60）を用いた固体型色素増感太陽電池に D102 を利用したものが 2005 年に報告されている[50]。1.6 μm の TiO_2 電極を用いた時，変換効率 4.1 ％を達成している。D102 はモル吸光係数が高いため，わずか 1.6 μm の TiO_2 電極であっても 450～550nm の光はほとんど吸収することが可能なため，今までにない効率が達成できたものと考えられる。N719 色素は 2.0 μm の TiO_2 電極でも約半分程度しか吸収することができない（図 61）。

図60 スピロ型ホール輸送材料の化学構造

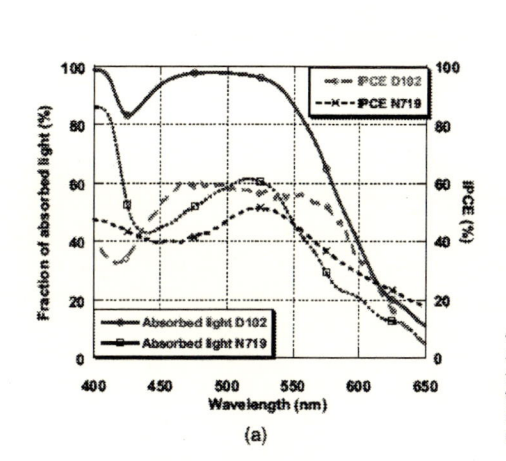

図61(a) 各色素の UV-Vis 吸収スペクトルと IPCE スペクトル

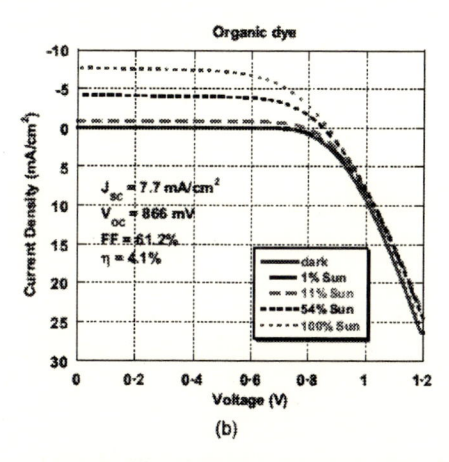

図61(b) D102 を用いた固体色素増感太陽電池の IV 特性

23 おわりに

2000年頃の日本で学会では「有機色素は Ru 錯体に遠く及ばず，絶対に置き換え不可能だと」断言されていた。しかしながら，個々の特性においてはブレークスルーが果たされており，もう少しのところで N719 を越えるレベルまで来ている。

筆者は今後も有機色素は大きな課題を次々にクリアしていくと期待している。特に最も大きな期待は IR 領域の光電変換特性である。これがクリアできれば，本当にシリコンと同等の変換効

率を達成することができるだろう。

文　献

1) J. Moser, Notizuber Verstarkung photoelektrischer Strome durch optische Sensibilisirung, Monatshefte fur Chemie, **8** (1), 373 (1887)
2) H. Tsubomura, M. Matsumura, Y. Nomura and T. Amamiya, Dye sensitised zinc oxide : aqueous electrolyte : platinum photocell., *Nature*, **261**, 402 (1976)
3) B. O'Regan and M. Grätzel, A low cost, high efficiency solar cell based on dye sensitized colloidal TiO_2 films, *Nature*, **353** (24), 737 (1991)
4) K. Sayama, M. Sugino, H. Sugihara, Y. Abe and H. Arakawa, Photosensitization of Porous TiO_2 Semiconductor Electrode with Xanthene Dyes, *Chem. Lett.*, 753 (1998)
5) K. Hara, T. Horiguchi, T. Kinoshita, K. Sayama, H. Sugihara and H. Arakawa, Highly Efficient Photon-to-Electron Conversion of Mercurochrome-sensitized Nanoporous ZnO Solar Cells, *Chem. Lett.*, 316 (2000)
6) S. Ferrere, A. Zaban and B. A. Gregg, Dye Sensitization of Nanocrystalline Tin Oxide by Perylene Derivatives, *J. Phys. Chem. B*, **101**, 4490 (1997)
7) 特開平 11-086916 号公報, 特開平 11-238925 号公報
8) A. C. Khazraji, S. Hotchandani, S. Das, and P. V. Kamat, Controlling Dye (Merocyanine-540) Aggregation on Nanostructured TiO_2 Films. An Organized Assembly Approach for Enhancing the Efficiency of Photosensitization, *J. Phys. Chem. B*, **103**, 4693 (1999)
9) K. Sayama, K. Hara, N. Mori, M. Satsuki, S. Suga, S. Tsukagoshi, Y. Abe, H. Sugihara and H. Arakawa, Photosensitization of a Porous TiO_2 Electrode with Merocyanine Dyes Containing a Carboxyl Group and a Long Alkyl Chain, *Chem. Commun.*, 1173 (2000)
10) K. Sayama, S. Tsukagoshi, K. Hara, Y. Ohga, A. Shinpou, Y. Abe, S. Suga and H. Arakawa, Photoelectrochemical Properties of J Aggregates of Benzothiazole Merocyanine Dyes on a Nanostructured TiO_2 Film, *J. Phys. Chem. B*, **106**, 1363 (2002)
11) Edwin. E. Jelley, Spectral Absorption and Fluorescene of Dyes in the Molecular State, *Nature*, **138**, 1009 (1936)
12) F. Nulesch, J. E. Moser, V. Shklover, and M. Grätzel, Merocyanine Aggregation in Mesoporous Networks, *J. Am. Chem. Soc.*, **118**, 5420 (1996)
13) 特開平 11-086916 号公報, 特開平 11-214730 号公報
14) A. Ehret, L. Stuhl and M. T. Spitler, Variation of carboxylate-functionalized cyanine dyes to produce efficient spectral sensitization of nanocrystalline solar cells, *Electrochim. Acta*, **45**, 4553 (2000)
15) K. Sayama, K. Hara, Y. Ohga, A. Shinpou, S. Suga and H. Arakawa, Significant efects of the distance between the cyanine dye skeleton and the semiconductor surface on the

photoelectrochemical properties of dye-sensitized porous semiconductor electrodes, *New. J. Chem.*, **25**, 200 (2001)

16) W. Zhao, Y. J. Hou, X. S. Wang, B. W. Zhang, Y. Cao, R. Yang, W. B. Wang and X. R. Xiao, Study on squarylium cyanine dyes for photoelectric conversion, *Sol. Ener. Mater. Sol. Cells*, **58**, 173 (1999)

17) A. Burke, L. Schmidt-Mende, S. Ito and M. Grätzel, A novel blue dye for near-IR dye-sensitised solar cell applications, *Chem. Commun.*, 234 (2007)

18) O. Enea, J. Moser and M. Grätzel, Achievement of incident photon to electric current conversion yields exceeding 80% in the spectral sensitization of titanium dioxide by coumarin, *J. Electroanal. Chem.*, **259**, 59 (1989)

19) K. Hara, K. Sayama, Y. Ohga, A. Shinpo, S. Suga and H. Arakawa, A coumarin-derivative dye sensitized nanocrystalline TiO_2 solar cell having a high solar-energy conversion efficiency up to 5.6% *Chem. Commun.*, 569 (2001)

20) K. Hara, M. Kurashige, Y. Dan-oh, C. Kasada, A. Shinpo, S. Suga, K. Sayama and H. Arakawa, *New. J. Chem.*, **27**, 783 (2003)

21) 平成15年度㈱新エネルギー・産業技術総合開発機構受託研究（委託業務），成果報告書，太陽光発電技術研究開発，革新的次世代太陽光発電システム技術研究開発「高性能色素増感太陽電池技術の研究開発」，P151, 平成16年3月

22) Z. Wang, Y. Cui, Y. Dan-oh, C. Kasada, A. Shinpo and K. Hara, Thiophene-Functionalized Coumarin Dye for Efficient Dye-Sensitized Solar Cells : Electron Lifetime Improved by Coadsorption of Deoxycholic Acid, *J. Phys. Chem. C*, **111**, 7224 (2007)

23) K. Hara, M. Kurashige, S. Ito, A. Shinpo, S. Suga, K. Sayama and H. Arakawa, Novel polyene dyes for highly efficient dye-sensitized solar cells, *Chem. Commun.*, 252 (2003)

24) S. Kim, J. K. Lee, S. O. Kang, J. Ko, J.-H. Yum, S. Fantacci, F. D. Angelis, D. D. Censo, Md. K. Nazeeruddin and M. Grätzel, Molecular Engineering of Organic Sensitizers for Solar Cell Applications, *J. Am. Chem. Soc.*, **128**, 16701 (2006)

25) S. Hwang, J. H. Lee, C. Park, H. Lee, C. Kim, C. Park, Mi-H. Lee, W. Lee, J. Park, K. Kim, N-G. Park and C. Kim, A highly efficient organic sensitizer for dye-sensitized solar cells, *Chem. Commun.*, 4887 (2007)

26) N. Koumura, Z-S. Wang, S. Mori, M. Miyashita, E. Suzuki and K. Hara, Alkyl-Functionalized Organic Dyes for Efficient Molecular Photovoltaics, *J. Am. Chem. Soc.*, **128**, 14256 (2006)

27) Z-S. Wang, N. Koumura, Y. Cui, M. Takahashi, H. Sekiguchi, A. Mori, T. Kubo, A. Furube, and K. Hara, Hexylthiophene-Functionalized Carbazole Dyes for Efficient Molecular Photovoltaics : Tuning of Solar-Cell Performance by Structural Modification, *Chem. Mater.*, **20**, 3993 (2008)

28) K. Kalyanasundaram, N. Vlachopoulos, V. Krishnan, A. Monnier,t and M. Grätzel, Sensitlzation of TiO, in the Visible Light Region Using Zinc Porphyrins, *J. Phys. Chem.*, **91**, 2342 (1987)

29) A. Kay and M. Grätzel, Artificial photosynthesis. 1. Photosensitization of titania solar

cells with chlorophyll derivatives and related natural porphyrins, *J. Phys. Chem. B*, **97**, 6272 (1993)

30) T. Ma, K. Inoue, K. Yao, H. Noma, T. Shuji, E. Abe, J. Yu, X. Wang, and B. Zhang, *J. Electroanal. Chem.*, **537**, 31, (2002)

31) 特開 2006-032260 号公報

32) F.-R. F. Fan, A. J. Bard, Spectral sensitization of the heterogeneous photocatalytic oxidation of hydroquinone in aqueous solutions at phthalocyanine-coated titanium dioxide powders, *J. Am. Chem. Soc.*, **101**, 6319 (1979)

33) A. Giraudeau, F-R. F. Fan and A. J. Bard, Semiconductor electrodes. 30. Spectral sensitization of the semiconductors titanium oxide (n-TiO_2) and tungsten oxide (n-WO_3) with metal phthalocyanines, *J. Am. Chem. Soc.*, **102**, 5137 (1980)

34) Y-C. Shen, L. Wang, Z. Lu, Y. Wei, Q. Zhou, H. Mao and H. Xu, Fabrication, characterization and photovoltaic study of a TiO_2 microporous electrode, *Thin Solid Films*, **257**, 144 (1995)

35) J. Fang, H. Mao, J. Wu, X. Zhang and Z. Lu, The photovoltaic study of co-sensitized microporous TiO2 electrode with porphyrin and phthalocyanine molecules, *Appl. Surf. Sci.*, **119**, 237 (1997)

36) M. K. Nazeeruddin, R. Humphry-Baker, M. Grätzel, D. Wöhrle, G. Schnurpfeil, G. Schneider, A. Hirth and N. Trombach, Efficient near-IR sensitization of nanocrystalline TiO_2 films by zinc and aluminum phthalocyanines, *J. Porphyrins Phthalocyanines*, **3**, 230 (1999)

37) P. Y. Reddy, L. Giribabu, C. Lyness, H. J. Snaith, C. Vijaykumar, M. Chandrasekharam, M. Lakshmikantam, J-H Yum, K. Kalyanasundaram, M. Grätzel, M. K. Nazeeruddin, Efficient Sensitization of Nanocrystalline TiO_2 Films by a Near-IR-Absorbing Unsymmetrical Zinc Phthalocyanine, *Angew. Chem. Int. Ed.*, **46**, 373 (2007)

38) J-H. Yum, S-r. Jang, R. Humphry-Baker, M. Grätzel, J-J. Cid, T. Torres and Md. K. Nazeeruddin, Effect of Coadsorbent on the Photovoltaic Performance of Zinc Pthalocyanine-Sensitized Solar Cells, *Langumuir*, **24**, 5436 (2008)

39) H. Osora, W. Li, L. Otero and M. A. Fox, Photosensitization of nanocrystalline TiO_2 thin films by a polyimide bearing pendent substituted-Ru $(bpy)_3^{+2}$ groups, *J. Photochem. Photobiol. B*, **43**, 232 (1998)

40) G. K. R. Senadeera, K. Nakamura, T. Kitamura, Y. Wada and S. Yanagida, Fabrication of highly efficient polythiophene-sensitized metal oxide photovoltaic cells, *Appl. Phys. Lett.*, **83**, 5470 (2003)

41) a) 堀内保, 三浦偉俊, 内田聡「色素増感太陽電池の高効率メタルフリー増感色素」2003 年電気化学秋季大会 2L05, b) 内田聡「高効率色素増感型太陽電池の材料開発」2007 年 2 月 2 日静岡大学薄膜基板研究懇話会第 10 回記念研究発表会, c) S. Uchida, S. Ito, M. Takata, H. Miura, Progress in non-Ru dyes, 2nd INTERNATIONAL CONFERENCE on the INDUSTRIALISATION of DSC, 12th September, 2007

42) T. Horiuchi, H. Miura and S. Uchida, Highly-efficient metal-free organic dyes for dye-

sensitized solar cells, *Chem. Commun.*, 3036（2003）

43) T. Horiuchi, H. Miura and S. Uchida, Highly efficient metal-free organic dyes for dye-sensitized solar cells, *J. Photochem. Photobiol. A*, **164**, 29（2004）

44) T. Horiuchi, H. Miura, K. Sumioka and S. Uchida, High Efficiency of Dye-Sensitized Solar Cells Based on Metal-Free Indoline Dyes, *J. Am. Chem. Soc.*, **126**, 12218（2004）

45) S. Ito, S. M. Zakeeruddin, R. Humphry-Baker, P. Liska, R. Charvet, P. Comte, M. K. Nazeeruddin, P. Péchy, M. Takata, H. Miura, S. Uchida, and M. Grätzel, High-Efficiency Organic-Dye-Sensitized Solar Cells Controlled by Nanocrystalline-TiO_2 Electrode Thickness, *Adv. Mater.*, **18**, 1202（2006）

46) D. Kuang, S. Uchida, R. Humphry-Baker, S. M. Zakeeruddin and M. Grätzel, Organic Dye-Sensitized Ionic Liquid Based Solar Cells：Remarkable Enhancement in Performance through Molecular Design of Indoline Sensitizers, *Angew. Chem. Int. Ed.*, **47**, 1923（2008）

47) S. Ito, submitted

48) 長谷川仁, 内田聡, 「ZnO内包複合型酸化物電極を用いた色素増感太陽電池」, 2003年 電気化学秋季大会, 2L25（2003）

49) B. Onwona-Agyeman, S. Kaneko, A. Kumara, M. Okuya, K. Murakami, A. Konno and K. Tennakone, Sensitization of Nanocrystalline SnO_2 Films with Indoline Dyes, *Jpn. J. Appl. Phys.*, **44**, L731（2005）

50) L. Schmidt-Mende, U. Bach, R. Humphry-Baker, T. Horiuchi, H. Miura, S. Ito, S. Uchida and, M. Grätzel, Organic Dye for Highly Efficient Solid-State Dye-Semsitized Solar Cells, *Adv. Mater.*, **17**, 813（2005）

色素データ編

Name	Anthocyanine
IUPAC Name	
Type	
ε /Mol$^{-1}\cdot$cm^{-1}	λ_{max}/nm
HOMO/V vs SCE	LUMO/V vs SCE
Structure	$C_{21}H_{20}O_{11}$ Exact Mass : 448.10 Mol. Wt. : 448.38 C, 56.25 ; H, 4.50 ; O, 39.25
ref.	N. J. Cherepy et. al., J. Phys. Chem. B, 101, 9342 (1997)
lab.	Grätzel
remark	

色素増感太陽電池研究者のための色素データ集

Name	N3
IUPAC Name	*cis*-di(thiocyanato)-bis(2,2'-bipyridyl-4,4'-dicarboxylic acid)-ruthenium(Ⅱ)
Type	Ru complex
ε /Mol^{-1}·cm^{-1}	λ_{max}/nm
48200 EtOH	at 314
14000 EtOH	at 389
14200 EtOH	at 534
13900 t-BuOH／AcN (1/1)	at 541
HOMO/V vs SCE	LUMO/V vs SCE
Structure	C$_{26}$H$_{16}$N$_6$O$_8$RuS$_2$ Exact Mass：705.95 Mol. Wt.：705.64 C, 44.25；H, 2.29；N, 11.91；O, 18.14；Ru, 14.32；S, 9.09

ref.	Brian O'Regan, Michael Grätzel A low-cost, high-efficiency solar cell based on dye-sensitized colloidal TiO$_2$ films Nature, 353, 737 (1991) M. K. Nazeeruddin, A. Kay, I. Rodicio, R. Humphry-Baker, E. Mueller, P. Liska, N. Vlachopoulos, M. Graetzel Conversion of light to electricity by cis-X2bis (2,2'-bipyridyl-4,4'-dicarboxylate) ruthenium (II) charge-transfer sensitizers (X=Cl-, Br-, I-, CN-, and SCN-) on nanocrystalline titanium dioxide electrodes J. Am. Chem. Soc., 115, 6382 (1993) Md. K. Nazeeruddin, S. M. Zakeeruddin, R. Humphry-Baker, M. Jirousek, P. Liska, N. Vlachopoulos, V. Shklover, Christian-H. Fischer, M. Grätzel Acid-Base Equilibria of (2,2'-Bipyridyl-4,4'-dicarboxylic acid)ruthenium(II) Complexes and the Effect of Protonation on Charge-Transfer Sensitization of Nanocrystalline Titania Inorg. Chem., 38, 6298 (1999)
lab.	Grätzel
remark	violet/solution, red/on TiO$_2$

色素増感太陽電池研究者のための色素データ集

Name	N621
IUPAC Name	*cis*-di(thiocyanato)-(2,2'-bipyridyl-4,4'-dicarboxylic acid)(4,4'-ditridecyl-2,2'-bipyridyl)-ruthenium(II)
Type	Ru complex
ε /Mol^{-1}·cm^{-1}	λ_{max}/nm
HOMO/V vs SCE	LUMO/V vs SCE
Structure	$C_{50}H_{68}N_6O_4RuS_2$ Exact Mass : 982.38 Mol. Wt. : 982.31 C, 61.13 ; H, 6.98 ; N, 8.56 ; O, 6.51 ; Ru, 10.29 ; S, 6.53
ref.	Mohammad K. Nazeeruddi, Filippo De Angeli, Simona Fantacci, Annabella Selloni, Guido Viscardi, Paul Liska, Seigo Ito, Bessho Takeru, Michael Grätzel Combined Experimental and DFT-TDDFT Computational Study of Photoelectrochemical Cell Ruthenium Sensitizers J. Am. Chem. Soc., 127, 16835 (2005) M. K. Nazeeruddin, D. Di Censo, R. Humphry-Baker, M. Grätzel Highly Selective and Reversible Optical, Colorimetric, and Electrochemical Detection of Mercury (II) by Amphiphilic Ruthenium Complexes Anchored onto Mesoporous Oxide Films Adv. Mater, 16, 189 (2006)
lab.	Grätzel
remark	

Name	N712
IUPAC Name	tetra(tetrabutylammonium)[cis-di(thiocyanato)-bis(2,2'-bipyridyl-4,4'-dicarboxylate)-ruthenium(II)]
Type	Ru complex
ε /Mol$^{-1}\cdot$cm^{-1}	λ_{max}/nm
HOMO/V vs SCE	LUMO/V vs SCE
Structure	$C_{90}H_{156}N_{10}O_8RuS_2$ Exact Mass : 1671.06 Mol. Wt. : 1671.46 C, 64.67 ; H, 9.41 ; N, 8.38 ; O, 7.66 ; Ru, 6.05 ; S, 3.84
ref.	Md. K. Nazeeruddin, R. Humphry-Baker, P. Liska, M. Grätzel Investigation of Sensitizer Adsorption and the Influence of Protons on Current and Voltage of a Dye-Sensitized Nanocrystalline TiO$_2$ Solar Cell J. Phys. Chem. B, 107, 8981 (2003)
lab.	Grätzel
remark	

色素増感太陽電池研究者のための色素データ集

Name	N719
IUPAC Name	bis(tetrabutylammonium ([*cis*-di(thiocyanato)-bis(2,2'-bipyridyl-4-carboxylate-4'-carboxylic acid)-ruthenium(Ⅱ)]
Type	Ru complex
ε /Mol$^{-1}\cdot$cm^{-1}	λ_{max}/nm
45900 EtOH	at 308
13300 EtOH	at 380
13000 EtOH	at 518
13600?	at 535
HOMO/V vs SCE	LUMO/V vs SCE
Structure	$C_{58}H_{86}N_8O_8RuS_2$ Exact Mass：1188.51 Mol. Wt.：1188.55 C, 58.61；H, 7.29；N, 9.43；O, 10.77；Ru, 8.50；S, 5.40

ref.	M. K. Nazeeruddin, F. D. Angelis, S. Fantacci, A. Selloni, G. Viscardi, P. Liska, S. Ito, B. Takeru, M. Grätzel Combined Experimental and DFT-TDDFT Computational Study of Photoelectrochemical Cell Ruthenium Sensitizers J. Am. Chem. Soc., 127, 16835 (2005)
lab.	Grätzel
remark	violet/solution, red/on TiO_2

色素増感太陽電池研究者のための色素データ集

Name	N749
IUPAC Name	(2,2':6',2"-terpyridine-4,4',4"-tricarboxylate)ruthenium(II)tris(tetrabutylammonium)tris(isothiocyanate)
Type	Ru complex

ε /Mol$^{-1}\cdot$cm^{-1}	λ_{max}/nm
6000 H$_2$O	at 570
7320 AcN	at 605
7480 EtOH	at 605
7900 DMF	at 609

HOMO/V vs SCE	LUMO/V vs SCE

Structure	$C_{69}H_{116}N_9O_6RuS_3$ Exact Mass: 1364.73 Mol. Wt.: 1364.98 C, 60.71; H, 8.57; N, 9.24; O, 7.03; Ru, 7.40; S, 7.05
ref.	Mohammad K. Nazeeruddin, Peter Péchy, Thierry Renouard, Shaik M. Zakeeruddin, Robin Humphry-Baker, Pascal Comte, Paul Liska, Le Cevey, Emiliana Costa, Valery Shklover, Leone Spiccia, Glen B. Deacon, Carlo A. Bignozzi, Michael Grätzel Engineering of Efficient Panchromatic Sensitizers for Nanocrystalline TiO$_2$-Based Solar Cells J. Am. Chem. Soc., 123 (8), 1613 (2001) Kohjiro Hara, Takeshi Nishikawa, Kazuhiro Sayama, Kenichi Aika, Hironori Arakawa Novel and Efficient Organic Liquid Electrolytes for Dye-sensitized Solar Cells Based on a Ru(II) Terpyridyl Complex Photosensitizer Chem. Lett., 32, 1014 (2003)
lab.	Grätzel
remark	Black/solution, black/on TiO$_2$

Name	N820
IUPAC Name	*cis*-di(thiocyanato)-(2,2'-bipyridyl-4,4'-dicarboxylic acid)(4,4'-dimethyl-2,2'-bipyridyl)-ruthenium(II)
Type	Ru complex
ε /Mol^{-1}·cm^{-1}	λ_{max}/nm
HOMO/V vs SCE	LUMO/V vs SCE
Structure	$C_{26}H_{20}N_6O_4RuS_2$ Exact Mass : 646.00 Mol. Wt. : 645.68 C, 48.36 ; H, 3.12 ; N, 13.02 ; O, 9.91 ; Ru, 15.65 ; S, 9.93
ref.	C. Lee, J.-H. Yum, H. Choi, S. O. Kang, J. Ko, R. H.umphry-Baker, M. Grätzel, M. K. Nazeeruddin Phenomenally High Molar Extinction Coefficient Sensitizer with "Donor-Acceptor" Ligands for Dye-Sensitized Solar Cell Applications Inorg. Chem., 47, 2267 (2008)
lab.	Grätzel
remark	

色素増感太陽電池研究者のための色素データ集

Name	N823	
IUPAC Name	*cis*-di(thiocyanato)-(2,2'-bipyridyl-4,4'-dicarboxylic acid)(4,4'-dihexyl-2,2'-bipyridyl)-ruthenium(II)	
Type	Ru complex	
ε /Mol^{-1}·cm^{-1}		λ_{max}/nm
HOMO/V vs SCE		LUMO/V vs SCE
Structure	$C_{36}H_{40}N_6O_4RuS_2$ Exact Mass : 786.16 Mol. Wt. : 785.94 C, 55.01 ; H, 5.13 ; N, 10.69 ; O, 8.14 ; Ru, 12.86 ; S, 8.16	
ref.	J.-J. Lagref, M. K. Nazeeruddin, M. Grätzel Molecular Engineering on Semiconductor Surfaces : Design, Synthesis and Application of New Efficient Amphiphilic Ruthenium Photosensitizers for Nanocrystalline TiO$_2$ Solar Cells Synth. Met., 138, 333 (2003) J. E. Kroeze, N. Hirata, S. Koops, M. K. Nazeeruddin, L. Schmidt-Mende, M. Grätzel, J. R. Durrant Alkyl Chain Barriers for Kinetic Optimization in Dye-Sensitized Solar Cells J. Am. Chem. Soc., 128, 16376 (2006)	
lab.	Grätzel	
remark		

Name	N845
IUPAC Name	*cis*-di(thiocyanato)-(2,2'-bipyridyl-4,4'-dicarboxylic acid)(4-[4-(*N*,*N*-di-*p*-anisylamino)phenoxymethyl]-4'-methyl-2,2'-bipyridyl)-ruthenium(II)
Type	Ru complex
ε /Mol^{-1}·cm^{-1}	λ_{max}/nm
49100DMF	at312
14300DMF	at396
14700DMF	at535
HOMO/V vs SCE	LUMO/V vs SCE
Structure	C$_{47}$H$_{41}$N$_7$O$_7$RuS$_2$ Exact Mass : 981.16 Mol. Wt. : 981.07 C, 57.54 ; H, 4.21 ; N, 9.99 ; O, 11.42 ; Ru, 10.30 ; S, 6.54
ref.	Narukuni Hirata, Jean-Jacques Lagref, Emilio J. Palomares, James R. Durrant, M. Khaja Nazeeruddin, Michael Grätzel, Davide Di Censo Supramolecular Control of Charge-Transfer Dynamics on Dye-sensitized Nanocrystalline TiO$_2$ Films Chem. Eur. J., 10, 595 (2004) Neil Robertson Optimizing Dyes for Dye-Sensitized Solar Cells Angew. Chem. Int. Ed., 45, 2338 (2006)
lab.	Grätzel
remark	

Name	N886
IUPAC Name	*trans*-[Ru(L)(NCS)2], L = 4,4'''-di-*tert*-butyl-4',4''-bis(carboxylic acid)-2,2' : 6',2'' : 6'',2'''-quaterpyridine
Type	Ru complex
ε /Mol^{-1}·cm^{-1}	λ_{max}/nm
HOMO/V vs SCE	LUMO/V vs SCE
Structure	$C_{32}H_{28}N_6O_4RuS_2$ Exact Mass：726.07 Mol. Wt.：725.80 C, 52.95；H, 3.89；N, 11.58；O, 8.82；Ru, 13.93；S, 8.84
ref.	C. Barolo, Md. K. Nazeeruddin, Simona Fantacci, D. Di Censo, P. Comte, P. Liska, G. Viscardi, P. Quagliotto, Filippo De Angelis, S. Ito, M. Grätzel Synthesis, Characterization, and DFT-TDDFT Computational Study of a Ruthenium Complex Containing a Functionalized Tetradentate Ligand Inorg. Chem., 45, 4642 (2006)
lab.	Grätzel
remark	

Name	N945
IUPAC Name	tetrabutylammonium [*cis*-di(thiocyanato)-(2,2'-bipyridyl-4-carboxylic acid-4'-carboxylate)(4,4'-di-(2-(3,6-dimethoxyphenyl)ethenyl)-2,2'-bipyridyl)-ruthenium(II)]
Type	Ru complex
ε /Mol$^{-1}\cdot$cm^{-1}		λ_{max}/nm	
HOMO/V vs SCE		LUMO/V vs SCE	
Structure	$C_{44}H_{36}N_6O_8RuS_2$ Exact Mass : 942.11 Mol. Wt. : 941.99 C, 56.10 ; H, 3.85 ; N, 8.92 ; O, 13.59 ; Ru, 10.73 ; S, 6.81
ref.	M. K. Nazeeruddin, T. Bessho, L. Cevey, S. Ito, C. Klein, F. D. Angelis, S. Fantacci, P. Comte, P. Liska, H. Imai, M. Graetzel A High Molar Extinction Coefficient Charge Transfer Sensitizer and Its Application in Dye-Sensitized Solar Cell J. Photochem. Photobiol., A, 185, 331 (2007)
lab.	Grätzel
remark	

Name	K9
IUPAC Name	*cis*-di(thiocyanato)-(2,2'-bipyridyl-4,4'-bis(carboxyvinyl))-(2,2'-bipyridine-4,4'-dinonyl)-ruthenium(Ⅱ)
Type	Ru complex
$\varepsilon /\mathrm{Mol}^{-1}\cdot\mathrm{cm}^{-1}$	$\lambda_{max}/\mathrm{nm}$
HOMO/V vs SCE	LUMO/V vs SCE
Structure	$C_{46}H_{56}N_6O_4RuS_2$ Exact Mass : 922.28 Mol. Wt. : 922.17 C, 59.91 ; H, 6.12 ; N, 9.11 ; O, 6.94 ; R, 10.96 ; S, 6.95
ref.	M. K. Nazeeruddin, C. Klein, P. Liska, M. Grätzel Synthesis of Novel Ruthenium Sensitizers and Their Application in Dye-Sensitized Solar Cells Coord. Chem. Rev., 249, 1460 (2005)
lab.	Grätzel
remark	

Name	K19
IUPAC Name	*cis*-di(thiocyanato)-(2,2'-bipyridyl-4,4'-dicarboxylic acid)(4,4'-bis(*p*-hexyloxystyryl)-2,2'-bipyridyl)-ruthenium(II)
Type	Ru complex
ε /Mol^{-1}·cm^{-1}	λ_{max}/nm
18200 AcN/t-BuOH (1/1)	at 543
HOMO/V vs SCE	LUMO/V vs SCE

Structure	C$_{48}$H$_{44}$N$_6$O$_6$RuS$_2$ Exact Mass : 966.18 Mol. Wt. : 966.10 C, 59.67 ; H, 4.59 ; N, 8.70 ; O, 9.94 ; Ru, 10.46 ; S, 6.64
ref.	Peng Wang, Cédric Klein, Robin Humphry-Baker, Shaik M. Zakeeruddin, Michael Grätzel A High Molar Extinction Coefficient Sensitizer for Stable Dye-Sensitized Solar Cells J. Am. Chem. Soc., 127, 808 (2005) Daibin Kuang, Seigo Ito, Bernard Wenger, Cedric Klein, Jacques-E Moser, Robin Humphry-Baker, Shaik M. Zakeeruddin, Michael Grätzel High Molar Extinction Coefficient Heteroleptic Ruthenium Complexes for Thin Film Dye-Sensitized Solar Cells J. Am. Chem. Soc., 128, 4146 (2006)
lab.	Grätzel
remark	

色素増感太陽電池研究者のための色素データ集

Name	K23
IUPAC Name	
Type	Ru complex
ε /Mol^{-1}·cm^{-1}	λ_{max}/nm
HOMO/V vs SCE	LUMO/V vs SCE
Structure	
ref.	
lab.	Grätzel
remark	

Name	K51
IUPAC Name	sodium[*cis*-di(thiocyanato)-(2,2'-bipyridyl-4-carboxylic acid-4'-carboxylate)(4,4'-bis[(triethylene glycol methyl ether)methyl ether]-2,2'-bipyridyl)-ruthenium(II)]
Type	Ru complex

ε /Mol$^{-1}\cdot$cm^{-1}		λ_{max}/nm	

HOMO/V vs SCE		LUMO/V vs SCE	

Structure	
	C$_{40}$H$_{47}$N$_6$NaO$_{12}$RuS$_2$ Exact Mass : 992.16 Mol. Wt. : 992.03 C, 48.43 ; H, 4.78 ; N, 8.47 ; Na, 2.32 ; O, 19.35 ; Ru, 10.19 ; S, 6.46
ref.	Daibin Kuang, Cedric Klein, Henry J. Snaith, Jacques-E Moser, Robin Humphry-Baker, Pascal Comte, Shaik M. Zakeeruddin, Michael Grätzel Ion Coordinating Sensitizer for High Efficiency Mesoscopic Dye-Sensitized Solar Cells : Influence of Lithium Ions on the Photovoltaic Performance of Liquid and Solid-State Cells Nano Lett., 6 (4), 769 (2006)
lab.	Grätzel
remark	8.10% vs 6.73%(Z-907)

Name	K60
IUPAC Name	*cis*-di(thiocyanato)-(2,2'-bipyridyl-4,4'-dicarboxylic acid)(4,4'-bis(2-(4-(1,4,7,10-tetraoxyundecyl)phenyl)ethenyl)-2,2'-bipyridyl)-ruthenium(II)
Type	Ru complex

ε /Mol^{-1}·cm^{-1}	λ_{max}/nm

HOMO/V vs SCE	LUMO/V vs SCE

Structure	$C_{54}H_{56}N_6O_{12}RuS_2$ Exact Mass : 1146.24 Mol. Wt. : 1146.26 C, 56.58 ; H, 4.92 ; N, 7.33 ; O, 16.75 ; Ru, 8.82 ; S, 5.59
ref.	D. Kuang, C. Klein, S. Ito, J.-E. Moser, R. Humphry-Baker, S. M. Zakeeruddin, M. Grätzel High Molar Extinction Coefficient Ion-Coordinating Ruthenium Sensitizer for Efficient and Stable Mesoscopic Dye-Sensitized Solar Cells Adv. Funct. Mater., 17, 154 (2007)
lab.	Grätzel
remark	

Name	K66		
IUPAC Name			
Type	Ru complex		
ε /Mol$^{-1}\cdot$cm^{-1}		λ_{max}/nm	
HOMO/V vs SCE		LUMO/V vs SCE	
Structure			
ref.	Grätzel		
lab.			
remark			

色素増感太陽電池研究者のための色素データ集

Name	K69
IUPAC Name	
Type	Ru complex
ε /Mol$^{-1}\cdot$cm^{-1}	λ_{max}/nm
HOMO/V vs SCE	LUMO/V vs SCE
Structure	
ref.	
lab.	Grätzel
remark	

Name	K73
IUPAC Name	*cis*-di(thiocyanato)-(2,2'-bipyridyl-4,4'-dicarboxylic acid)(4,4'-bis(*p*-methoxystyryl)-2,2'-bipyridyl)-ruthenium(Ⅱ)
Type	Ru complex

ε /Mol^{-1}·cm^{-1}		λ_{max}/nm	

HOMO/V vs SCE		LUMO/V vs SCE	

Structure

$C_{42}H_{32}N_6O_6RuS_2$
Exact Mass : 882.09
Mol. Wt. : 881.94
C, 57.20 ; H, 3.66 ; N, 9.53 ; O, 10.88 ; Ru, 11.46 ; S, 7.27

ref.	Daibin Kuang, Seigo Ito, Bernard Wenger, Cedric Klein, Jacques-E Moser, Robin Humphry-Baker, Shaik M. Zakeeruddin, Michael Grätzel High Molar Extinction Coefficient Heteroleptic Ruthenium Complexes for Thin Film Dye-Sensitized Solar Cells J. Am. Chem. Soc., 128, 4146 (2006)
lab.	Grätzel
remark	

色素増感太陽電池研究者のための色素データ集

Name	K77
IUPAC Name	*cis*-di(thiocyanato)-(2,2'-bipyridyl-4,4'-dicarboxylic acid)(4,4'-bis(2-(4-*tert*-butyloxyphenyl)ethenyl)-2,2'-bipyridyl)-ruthenium(Ⅱ)
Type	Ru complex
ε /Mol^{-1}·cm^{-1}	λ_{max}/nm
HOMO/V vs SCE	LUMO/V vs SCE
Structure	C$_{48}$H$_{44}$N$_6$O$_6$RuS$_2$ Exact Mass : 966.18 Mol. Wt. : 966.10 C, 59.67 ; H, 4.59 ; N, 8.70 ; O, 9.94 ; Ru, 10.46 ; S, 6.64

ref.	D. Kuang, C. Klein, S. Ito, J.-E. Moser, R. Humphry-Baker, N. Evans, F. Duriaux, C. Grätzel, S. M. Zakeeruddin, M. Grätzel High-Efficiency and Stable Mesoscopic Dye-Sensitized Solar Cells Based on a High Molar Extinction Coefficient Ruthenium Sensitizer and Nonvolatile Electrolyte Adv. Mater., 19 (8), 1133 (2007) D. Kuang et. al., JACS, 128 (24), 7732 (2006) D. Kuang, P. Wang, S. Ito, S. M. Zakeeruddin, M. Grätzel Stable Mesoscopic Dye-Sensitized Solar Cells Based on Tetracyanoborate Ionic Liquid Electrolyte J. Am. Chem. Soc., 128, 7732 (2006)
lab.	Grätzel
remark	highly stable

色素増感太陽電池研究者のための色素データ集

Name	Z316	
IUPAC Name	*cis*-di(thiocyanato)-(2,2'-bipyridyl-4,4'-dicarboxylic acid)(4-methyl-4'-hexadecyl-2,2'-bipyridyl)-ruthenium(Ⅱ)	
Type	Ru complex	
ε /Mol^{-1}·cm^{-1}		λ_{max}/nm
HOMO/V vs SCE		LUMO/V vs SCE
Structure	$C_{41}H_{50}N_6O_4RuS_2$ Exact Mass : 856.24 Mol. Wt. : 856.07 C, 57.52 ; H, 5.89 ; N, 9.82 ; O, 7.48 ; Ru, 11.81 ; S, 7.49	
ref.	H. Nusbaumer, J.-E. Moser, S. M. Zakeeruddin, M. K. Nazeeruddin, M. Grätzel CoII(dbbip)$_2^{2+}$ Complex Rivals Tri-iodide/Iodide Redox Mediator in Dye-Sensitized Photovoltaic Cells J. Phys. Chem. B 105, 10461 (2001)	
lab.	Grätzel	
remark		

Name	Z907
IUPAC Name	*cis*-di(thiocyanato)-(2,2'-bipyridyl-4,4'-dicarboxylic acid)(4,4'-dinonyl-2,2'-bipyridyl)-ruthenium(II)
Type	Ru complex

ε /Mol^{-1}·cm^{-1}	λ_{max}/nm
36500EtOH	at296
25300EtOH	at314
10000EtOH	at384
10100EtOH	at528
12200AcN/t-BuOH (1/1) ?	at 543?
HOMO/V vs SCE	LUMO/V vs SCE

Structure:

$C_{42}H_{52}N_6O_4RuS_2$
Exact Mass : 870.25
Mol. Wt. : 870.10
C, 57.98 ; H, 6.02 ; N, 9.66 ; O, 7.36 ; Ru, 11.62 ; S, 7.37

ref.	S. M. Zakeeruddin, Md. K. Nazeeruddin, R. Humphry-Baker, P. Péchy, P. Quagliotto, C. Barolo, G. Viscardi, M. Grätzel Design, Synthesis, and Application of Amphiphilic Ruthenium Polypyridyl Photosensitizers in Solar Cells Based on Nanocrystalline TiO_2 Films Langmuir, 18, 952 (2002) Peng Wang, Shaik M. Zakeeruddin, Jacques E. Moser, Mohammad K. Nazeeruddin, Takashi Sekiguchi, Michael Grätzel A stable quasi-solid-state dye-sensitized solar cell with an amphiphilic ruthenium sensitizer and polymer gel electrolyte Nature materials, 2, 402 (2003) Peng. Wang, S.M. Zakeeruddin, R. Humphry-Baker, J.E. Moser, M. Grätzel Molecular-Scale Interface Engineering of TiO_2 Nanocrystals : Improve the Efficiency and Stability of Dye-Sensitized Solar Cells Adv. Mater., 15, 2101 (2003)
lab.	Grätzel
remark	

Name	Z907Na
IUPAC Name	sodium[*cis*-di(thiocyanato)-(2,2'-bipyridyl-4-carboxylic acid-4'-carboxylate)(4,4'-dinonyl-2,2'-bipyridyl)-ruthenium(II)]
Type	Ru complex
ε /Mol^{-1}·cm^{-1}		λ_{max}/nm	
HOMO/V vs SCE		LUMO/V vs SCE	
Structure	$C_{42}H_{51}N_6NaO_4RuS_2$ Exact Mass : 892.24 Mol. Wt. : 892.08 C, 56.55 ; H, 5.76 ; N, 9.42 ; Na, 2.58 ; O, 7.17 ; Ru, 11.33 ; S, 7.19
ref.	Daibin Kuang, Peng Wang, Seigo Ito, Shaik. M. Zakeeruddin, Michael Grätzel Stable Mesoscopic Dye-Sensitized Solar Cells Based on Tetracyanoborate Ionic Liquid Electrolyte J. Am. Chem. Soc., 128, 7732 (2006)
lab.	Grätzel
remark	

色素増感太陽電池研究者のための色素データ集

Name	Z910	
IUPAC Name	*cis*-di(thiocyanato)-(2,2'-bipyridyl-4,4'-dicarboxylic acid)(4,4'-di(3-methoxystyryl)-2,2'-bipyridyl)-ruthenium(II)	
Type	Ru complex	
ε /Mol^{-1}·cm^{-1}		λ_{max}/nm
17010AcN 16850AcN		at410 at543
HOMO/V vs SCE		LUMO/V vs SCE
Structure	$C_{42}H_{32}N_6O_6RuS_2$ Exact Mass : 882.09 Mol. Wt. : 881.94 C, 57.20 ; H, 3.66 ; N, 9.53 ; O, 10.88 ; Ru, 11.46 ; S, 7.27	
ref.	P. Wang, S. M. Zakeeruddin, J. E. Moser, R. Humphry-Baker, P. Comte, V. Aranyos, A. Hagfeldt, M. K. Nazeeruddin, M. Grätzel Stable New Sensitizer with Improved Light Harvesting for Nanocrystalline Dye-Sensitized Solar Cells Adv. Mater., 16, 1806 (2004)	
lab.	Grätzel	
remark		

Name	WMC217
IUPAC Name	
Type	Porphyrin
ε /Mol$^{-1}\cdot$cm^{-1}	λ_{max}/nm
HOMO/V vs SCE	LUMO/V vs SCE
Structure	$C_{56}H_{53}N_5O_2Zn$ Exact Mass : 891.35 Mol. Wt. : 893.46 C, 75.28 ; H, 5.98 ; N, 7.84 ; O, 3.58 ; Zn, 7.32
ref.	MRS 2006 Fall Meeting, Boston
lab.	David L. Officer
remark	

色素増感太陽電池研究者のための色素データ集

Name	WMC234	
IUPAC Name		
Type	Porphyrin	
ε /Mol$^{-1}\cdot$cm^{-1}		λ_{max}/nm
HOMO/V vs SCE		LUMO/V vs SCE
Structure	C$_{66}$H$_{72}$N$_4$O$_4$Zn Exact Mass : 1048.48 Mol. Wt. : 1050.71 C, 75.44 ; H, 6.91 ; N, 5.33 ; O, 6.09 ; Zn, 6.23	
ref.	MRS 2006 Fall Meeting, Boston	
lab.	David L. Officer	
remark		

Name	WMC236
IUPAC Name	
Type	Porphyrin
ε /Mol^{-1}·cm^{-1}	λ_{max}/nm
HOMO/V vs SCE	LUMO/V vs SCE
Structure	$C_{58}H_{56}N_4O_4Zn$ Exact Mass : 936.36 Mol. Wt. : 938.50 C, 74.23 ; H, 6.01 ; N, 5.97 ; O, 6.82 ; Zn, 6.97
ref.	MRS 2006 Fall Meeting, Boston
lab.	David L. Officer
remark	

色素増感太陽電池研究者のための色素データ集

Name	WMC239		
IUPAC Name			
Type	Porphyrin		
ε /Mol$^{-1}\cdot$cm^{-1}		λ_{max}/nm	
HOMO/V vs SCE		LUMO/V vs SCE	
Structure	$C_{64}H_{70}N_4O_4Zn$ Exact Mass：1022.47 Mol. Wt.：1024.68 C, 75.02；H, 6.89；N, 5.47；O, 6.25；Zn, 6.38		
ref.	MRS 2006 Fall Meeting, Boston		
lab.	David L. Officer		
remark			

Name	WMC273
IUPAC Name	
Type	Porphyrin

ε /Mol^{-1}·cm^{-1}		λ_{max}/nm	
HOMO/V vs SCE		LUMO/V vs SCE	

Structure	C$_{62}$H$_{64}$N$_4$O$_4$Zn Exact Mass : 992.42 Mol. Wt. : 994.61 C, 74.87 ; H, 6.49 ; N, 5.63 ; O, 6.43 ; Zn, 6.58
ref.	Wayne M. Campbell, Kenneth W. Jolley, PaWel Wagner, Klaudia Wagner, Penny J. Walsh, Keith C. Gordon, Lukas Schmidt-Mende, Mohammad K. Nazeeruddin, Qing Wang, Michael Grätzel and David L. Officer, *J. Phys. Chem. C*, **111**, 11760-11762 (2007)
lab.	David L. Officer
remark	7.1%

色素増感太陽電池研究者のための色素データ集

Name	CYC-B1
IUPAC Name	*cis*-di(thiocyanato)-(2,2'-bipyridyl-4,4'-dicarboxylic acid)(4,4'-bis(5'-octyl-5-(2,2'-bithienyl))bipyridyl)-ruthenium(II)
Type	Ru complex

ε /Mol$^{-1}\cdot$cm^{-1}	λ_{max}/nm
35800DMF	at312
46400DMF	at400
21200DMF	at553

HOMO/V vs SCE	LUMO/V vs SCE
5.10	3.52

Structure:

$C_{56}H_{56}N_6O_4RuS_6$
Exact Mass : 1170.17
Mol. Wt. : 1170.55
C, 57.46 ; H, 4.82 ; N, 7.18 ; O, 5.47 ; Ru, 8.63 ; S, 16.44

ref.	Chia-Yuan Chen, Shi-Jhang Wu, Chun-Guey Wu, Jian-Ging Chen, Kuo-Chuan Ho A Ruthenium Complex with Superhigh Light-Harvesting Capacity for Dye-Sensitized Solar Cells Angew. Chem. Int. Ed., 45, 5822 (2006)
lab.	Chun-Guey Wu
remark	8.54% vs 7.70%(N3)

Name	Perylene
IUPAC Name	PPDCA perylene-3,4-dicarboxylic acid-9,10-(5-phenanthroline)carboximide PTCA perylene-3,4,9,10-tetracarboxylic acid BBAPDC N,N'-bis(2-benzoic acid)-3,4,9,10-perylenebis(dicarboximide)
Type	PPDCA PTCA BBDBAC
ε /Mol$^{-1}\cdot$cm^{-1}	λ_{max}/nm
HOMO/V vs SCE	LUMO/V vs SCE
Structure	PPDCA / PTCA / BBAPDC structures
ref.	Ferrere et. al., J. Phys. Chem. B, 101, 4490 (1997)
lab.	Gregg
remark	

色素増感太陽電池研究者のための色素データ集

Name	MKX-?
IUPAC Name	Mc[18,1] 3-carboxymethyl-5-[2-(3-octadecyl-2-benzothiazolinyldene)ethylidene]-2-thioxo-4-thiazolidinone Mc[2,1]（追加で分子軌道計算した色素） 3-carboxymethyl-5-[2-(3-ethyl-2-benzothiazolinyldene)ethylidene]-2-thioxo-4-thiazolidinone
Type	merocyanine
$\varepsilon / \text{Mol}^{-1} \cdot \text{cm}^{-1}$	λ_{max}/nm
HOMO/V vs SCE	LUMO/V vs SCE
Structure	$C_{32}H_{46}N_2O_3S_3$ Exact Mass：602.27 Mol. Wt.：602.91 C, 63.75；H, 7.69；N, 4.65；O, 7.96；S, 15.96
ref.	Kazuhiro Sayama, Kohjiro Hara, Hideki Sugihara, Hironori Arakawa, Nahoko Mori, Makoto Satsuki, Sadaharu Suga, Shingo Tsukagoshi, Yoshimoto Abe Photosensitization of a porous TiO_2 electrode with merocyanine dyes containing a carboxyl group and a long alkyl chain Chem. Commun., 1173 (2000) Kazuhiro Sayama, Shingo Tsukagoshi, Kohjiro Hara, Yasuyo Ohga, Akira Shinpou, Yoshimoto Abe, Sadaharu Suga, Hironori Arakawa Photoelectrochemical Properties of J Aggregates of Benzothiazole Merocyanine Dyes on a Nanostructured TiO_2 Film J. Phys. Chem. B, 106, 1363 (2002)
lab.	Hayashibara & AIST
remark	

Name	NKX-2311
IUPAC Name	2-cyano-5-(1,1,6,6-tetramethyl-10-oxo-2,3,5,6-tetrahydro-1H,4H,10H-11-oxa-3a-aza-benzo[de]anthracen-9-yl)-penta-2,4-dienoic acid
Type	coumarin
ε /Mol$^{-1}\cdot$cm^{-1}	λ_{max}/nm
51900 MeOH	at 504
HOMO/V vs SCE	LUMO/V vs SCE
1.28 solution	−0.82 solution.
Structure	$C_{25}H_{26}N_2O_4$ Exact Mass : 418.19 Mol. Wt. : 418.49 C, 71.75 ; H, 6.26 ; N, 6.69 ; O, 15.29
ref.	Kohjiro Hara, Kazuhiro Sayama, Hironori Arakawa, Yasuyo Ohga, Akira Shinpo, Sadaharu Suga A coumarin-derivative dye sensitized nanocrystalline TiO$_2$ solar cell having a high solar-energy conversion efficiency up to 5.6% Chem. Commun., 569 (2001)
lab.	Hayashibara & AIST
remark	red/solution, red/on TiO$_2$

色素増感太陽電池研究者のための色素データ集

Name	NKX-2510
IUPAC Name	2-cyano-5-(7-diethylamino-2-oxo-2H-chromen-3-yl)-penta-2,4-dienoic acid
Type	coumarin
ε /Mol$^{-1}\cdot$cm^{-1}	λ_{max}/nm
49800 MeOH	at 480
HOMO/V vs SCE	LUMO/V vs SCE
1.42 solution	−0.82 solution
Structure	$C_{19}H_{18}N_2O_4$ Exact Mass : 338.13 Mol. Wt. : 338.36 C, 67.44 ; H, 5.36 ; N, 8.28 ; O, 18.91
ref.	Kohjiro Hara, Tadatake Sato, Ryuzi Katoh, Akihiro Furube, Yasuyo Ohga, Akira Shinpo, Sadaharu Suga, Kazuhiro Sayama, Hideki Sugihara, Hironori Arakawa Molecular Design of Coumarin Dyes for Efficient Dye-Sensitized Solar Cells J. Phys. Chem. B., 107, 597 (2003)
lab.	Hayashibara & AIST
remark	red/solution, red/on TiO$_2$

Name	NKX-2569

	3d
IUPAC Name	2-cyano-7,7-bis(4-dimethylaminophenyl)hepta-2,4,6-trienoic acid
	2-cyano-5-(4-N,N-dimethylanilino)-*trans-trans*-penta-2,4-dienoic acid
	3d
	2-cyano-7,7-bis(4-diethylaminophenyl)hepta-2,4,6-trienoic acid
Type	polyene

ε /Mol$^{-1}\cdot$cm^{-1}	λ_{max}/nm
39400 t-BuOH : AN 50 : 50	at 501
HOMO/V vs SCE	LUMO/V vs SCE
0.87 solution	−1.1 solution

Structure	**(Structure 1)** C$_{24}$H$_{25}$N$_3$O$_2$ Exact Mass : 387.19 Mol. Wt. : 387.47 C, 74.39 ; H, 6.50 ; N, 10.84 ; O, 8.26 -------------------- **(Structure 2)** C$_{28}$H$_{33}$N$_3$O$_2$ Exact Mass : 443.26 Mol. Wt. : 443.58 C, 75.81 ; H, 7.50 ; N, 9.47 ; O, 7.21
ref.	Kohjiro Hara, Mitsuhiko Kurashige, Shunichiro Ito, Akira Shinpo, Sadaharu Suga, Kazuhiro Sayama, Hironori Arakawa Novel polyene dyes for highly efficient dye-sensitized solar cells Chem. Commun., 252 (2003) -------------------- Takayuki Kitamura, Masaaki Ikeda, Koichiro Shigaki, Teruhisa Inoue, Neil A. Anderson, Xin Ai, Tianquan Lian, Shozo Yanagida Phenyl-Conjugated Oligoene Sensitizers for TiO$_2$ Solar Cells Chem. Mat., 16, 1806 (2004)
lab.	Hayashibara & AIST -------------------- Nippon kayaku & Yanagida
remark	wine red/solution, red/on TiO$_2$

Name	NKX-2586
IUPAC Name	2-cyano-7-(1,1,6,6-tetramethyl-10-oxo-2,3,5,6-tetrahydro-1H,4H, 10H-11-oxa-3a-aza-benzo[de]anthracen-9-yl)-hepta-2,4,6-trienoic acid
Type	coumarin-methine

ε /Mol$^{-1}\cdot$cm^{-1}	λ_{max}/nm
59100 MeOH	at 506
HOMO/V vs SCE	LUMO/V vs SCE
1.15 solution	−0.83 solution

Structure	$C_{27}H_{28}N_2O_4$ Exact Mass : 444.20 Mol. Wt. : 444.52 C, 72.95 ; H, 6.35 ; N, 6.30 ; O, 14.40
ref.	Zhong-Sheng Wang, Kohjiro Hara, Yasufumi Dan-oh, Chiaki Kasada, Akira Shinpo, Sadaharu Suga, Hironori Arakawa, Hideki Sugihara Photophysical and (Photo) electrochemical Properties of a Coumarin Dye J. Phys. Chem. B, 109, 3907 (2005)
lab.	Hayashibara & AIST
remark	wine red/solution, violet/on TiO$_2$

色素増感太陽電池研究者のための色素データ集

Name	NKX-2587
IUPAC Name	2-cyano-3-[5-(1,1,6,6-tetramethyl-10-oxo-2,3,5,6-tetrahydro-1H,4H,10H-11-oxa-3a-aza-benzo[de]anthracen-9-yl)-thiophen-2-yl]-acrylic acid
Type	coumarin-thiophene
ε /Mol^{-1}·cm^{-1}	λ_{max}/nm
54300 t-BuOH : AN 50 : 50	at 507
HOMO/V vs SCE	LUMO/V vs SCE
1.01 on TiO$_2$	$-$0.85 on TiO$_2$
Structure	C$_{27}$H$_{26}$N$_2$O$_4$S Exact Mass : 474.16 Mol. Wt. : 474.57 C, 68.33 ; H, 5.52 ; N, 5.90 ; O, 13.49 ; S, 6.76
ref.	Kohjiro Hara, Zhong-Sheng Wang, Tadatake Sato, Akihiro Furube, Ryuzi Katoh, Hideki Sugihara, Yasufumi Dan-oh, Chiaki Kasada, Akira Shinpo, Sadaharu Suga Oligothiophene-Containing Coumarin Dyes for Efficient Dye-Sensitized Solar Cells J. Phys. Chem. B, 109, 15476 (2005)
lab.	Hayashibara & AIST
remark	red/solution, violet/on TiO$_2$

Name	NKX-2677
IUPAC Name	2-cyano-3-[5'-(1,1,6,6-tetramethyl-10-oxo-2,3,5,6-tetrahydro-1H,4H,10H-11-oxa-3a-aza-benzo[de]anthracen-9-yl)-[2,2']bithiophenyl-5-yl]-acrylic acid
Type	coumarin-thiophene
ε /Mol^{-1}·cm^{-1}	λ_{max}/nm
64300 t-BuOH : AN 50 : 50	at 511
HOMO/V vs SCE	LUMO/V vs SCE
0.93 on TiO$_2$	-0.89 on TiO$_2$
Structure	C$_{31}$H$_{28}$N$_2$O$_4$S$_2$ Exact Mass : 556.15 Mol. Wt. : 556.70 C, 66.88 ; H, 5.07 ; N, 5.03 ; O, 11.50 ; S, 11.52
ref.	Kohjiro Hara, Mitsuhiko Kurashige, Yasufumi Dan-oh, Chiaki Kasada, Akira Shinpo, Sadaharu Suga, Kazuhiro Sayama, Hironori Arakawa Design of new coumarin dyes having thiophene moieties for highly efficient organic-dye-sensitized solar cells New J. Chem., 27, 783 (2003) Kohjiro Hara, Zhong-Sheng Wang, Tadatake Sato, Akihiro Furube, Ryuzi Katoh, Hideki Sugihara, Yasufumi Dan-oh, Chiaki Kasada, Akira Shinpo, Sadaharu Suga Oligothiophene-Containing Coumarin Dyes for Efficient Dye-Sensitized Solar Cells J. Phys. Chem. B, 109, 15476 (2005)
lab.	Hayashibara & AIST
remark	wine red/solution, violet/on TiO$_2$

色素増感太陽電池研究者のための色素データ集

Name	NKX-2697	
IUPAC Name	2-cyano-3-[5″-(1,1,6,6-tetramethyl-10-oxo-2,3,5,6-tetrahydro-1H,4H,10H-11-oxa-3a-aza-benzo[de]anthracen-9-yl)-[2,2′;5′,2″]terthiophen-5-yl]-acrylic acid	
Type	coumarin-thiophene	
ε/Mol$^{-1}\cdot$cm^{-1}		λ_{max}/nm
73300 t-BuOH:AN 50:50		at 501
HOMO/V vs SCE		LUMO/V vs SCE
0.91 on TiO$_2$		-0.77 on TiO$_2$
Structure	$C_{35}H_{30}N_2O_4S_3$ Exact Mass : 638.14 Mol. Wt. : 638.82 C, 65.80 ; H, 4.73 ; N, 4.39 ; O, 10.02 ; S, 15.06	
ref.	Kohjiro Hara, Zhong-Sheng Wang, Tadatake Sato, Akihiro Furube, Ryuzi Katoh, Hideki Sugihara, Yasufumi Dan-oh, Chiaki Kasada, Akira Shinpo, Sadaharu Suga Oligothiophene-Containing Coumarin Dyes for Efficient Dye-Sensitized Solar Cells J. Phys. Chem. B, 109, 15476 (2005)	
lab.	Hayashibara & AIST	
remark	violet/solution, black/on TiO$_2$	

Name	NKX-?
IUPAC Name	(triethylammonium) {5-[2-(5-chloro-3-ethyl-3Hbenzothiazol-2-ylidene)-ethylidene]-3'-ethyl-4,4'-dioxo-2'-thioxo-[2,5']bithiazolidinyliden-3-yl} acetate
Type	merocyanine
ε /Mol$^{-1}\cdot$cm^{-1}	λ_{max}/nm
HOMO/V vs SCE	LUMO/V vs SCE
Structure	$C_{27}H_{33}ClN_4O_4S_4$ Exact Mass : 640.11 Mol. Wt. : 641.29 C, 50.57 ; H, 5.19 ; Cl, 5.53 ; N, 8.74 ; O, 9.98 ; S, 20.00
ref.	Hideo Otaka, Michie Kira, Kentaro Yano, Shunichiro Ito, Hirofumi Mitekura, Toshio Kawata, Fumio Matsui Multi-colored dye-sensitized solar cells J. Photochem. Photobiol. A Chem., 164, 67 (2004)
lab.	Hayashibara
remark	

色素増感太陽電池研究者のための色素データ集

Name	NKX-2753
IUPAC Name	cyano-{5,5-dimethyl-3-[2-(1,1,6,6-tetramethyl-10-oxo-2,3,5,6-tetrahydro-1H,4H,10H-11-oxa-3a-aza-benzo[de]anthracen-9-yl)vinyl]cyclohex-2-enylidene}-acetic acid
Type	coumarin-methine
ε /Mol$^{-1}\cdot$cm^{-1}	λ_{max}/nm
50300EtOH	at 492
HOMO/V vs SCE	LUMO/V vs SCE
0.90 on TiO$_2$	$-$0.87 on TiO$_2$
Structure	$C_{32}H_{36}N_2O_4$ Exact Mass : 512.27 Mol. Wt. : 512.64 C, 74.97 ; H, 7.08 ; N, 5.46 ; O, 12.48
ref.	Zhong-Sheng Wang, Kohjiro Hara, Yasufumi Dan-oh, Chiaki Kasada, Akira Shinpo, Sadaharu Suga, Hironori Arakawa, Hideki Sugihara Photophysical and (Photo) electrochemical Properties of a Coumarin Dye J. Phys. Chem. B, 109, 3907 (2005)
lab.	Hayashibara & AIST
remark	red/solution, violet/TiO$_2$

Name	NKX-2883
IUPAC Name	2-cyano-3-{5′-[1-cyano-2-(1,1,6,6-tetramethyl-10-oxo-2,3,5,6-tetrahydro-1H,4H,10H-11-oxa-3aaza-benzo[de]anthracen-9-yl)-vinyl]-[2,2′]bithiophenyl-5-yl}-acrylic acid
Type	coumarin-thiophene

ε /Mol^{-1}·cm^{-1}		λ_{max}/nm
97400EtOH		at 552
HOMO/V vs SCE		LUMO/V vs SCE
0.97 on TiO$_2$		−0.72 solution

Structure	$C_{34}H_{29}N_3O_4S_2$ Exact Mass : 607.16 Mol. Wt. : 607.74 C, 67.19 ; H, 4.81 ; N, 6.91 ; O, 10.53 ; S, 10.55
ref.	Zhong.-Sheng. Wang, Y. Cui, K. Hara, Y. Dan-oh, C. Kasada, A. Shinpo A High-Light-Harvesting-Efficiency Coumarin Dye for Stable Dye-Sensitized Solar Cells Adv. Mater., 19 (8), 1138 (2007)
lab.	Hayashibara & AIST
remark	violet/solution, black/on TiO$_2$

Name	Eosin Y
IUPAC Name	2',4',5',7'-Tetrabromofluorescein disodium salt
Type	xanthen
ε /Mol$^{-1}\cdot$cm^{-1}	λ_{max}/nm
HOMO/V vs SCE	LUMO/V vs SCE
Structure	$C_{20}H_8Br_4Na_2O_5$ Exact Mass : 689.69 Mol. Wt. : 693.87 C, 34.62 ; H, 1.16 ; Br, 46.06 ; Na, 6.63 ; O, 11.53
ref.	Kazuhiro Sayama, Maki Sugino, Hideki Sugihara, Yoshimoto Abe, Hironori Arakawa Photosensitization of Porous TiO$_2$ Semiconductor Electrode with Xanthene Dyes Chem. Lett., 753 (1998)
lab.	AIST
remark	

Name	Mercurochrome
IUPAC Name	
Type	xanthen
ε /Mol^{-1}·cm^{-1}	λ_{max}/nm
74000 EtOH	at 517
HOMO/V vs SCE	LUMO/V vs SCE
0.84 solution	-1.20 solution
Structure	(structure of mercurochrome) $C_{20}H_{10}Br_2HgNa_2O_6$ Exact Mass : 751.83 Mol. Wt. : 752.67 C, 31.92 ; H, 1.34 ; Br, 21.23 ; Hg, 26.65 ; Na, 6.11 ; O, 12.75
ref.	Kohjiro Hara, Takaro Horiguchi, Tohru Kinoshita, Kazuhiro Sayama, Hideki Sugihara, Hironori Arakawa Highly Efficient Photon-to-Electron Conversion of Mercurochrome-sensitized Nanoporous ZnO Solar Cells Chem. Lett., 316 (2000)
lab.	AIST
remark	orange/solution, red/on TiO$_2$

色素増感太陽電池研究者のための色素データ集

Name	MK-2
IUPAC Name	2-cyano-3-[5'''-(9-ethyl-9H-carbazol-3-yl)-3',3'',3''',4-tetra-n-hexyl-[2,2',5',2'',5'',2''']-quater-thiophenyl-5-yl]acrylic acid
Type	carbazole

$\varepsilon / \text{Mol}^{-1} \cdot \text{cm}^{-1}$	λ_{max}/nm
38400 THF : toluene 2 : 8	at 480
HOMO/V vs SCE	LUMO/V vs SCE
0.96 on TiO_2	-0.89 on TiO_2

Structure	$C_{58}H_{72}N_2O_2S_4$ Exact Mass : 956.45 Mol. Wt. : 957.46 C, 72.76 ; H, 7.58 ; N, 2.93 ; O, 3.34 ; S, 13.40
ref.	Nagatoshi Koumura, Zhong-Sheng Wang, Shogo Mori, Masanori Miyashita, Eiji Suzuki, Kohjiro Hara Alkyl-Functionalized Organic Dyes for Efficient Molecular Photovoltaics J. Am. Chem. Soc., 128 (44), 14256 (2006)
lab.	AIST
remark	8.10% red/solution, violet/on TiO_2

Name	D77
IUPAC Name	{5-[1,2,3,3a,4,8b-hexahydro-4-(4-methoxyphenyl)-cyclopenta[b]indole-7-ylmethylene]-4-oxo-2-thioxo-thiazolidin-3-yl} acetic acid
Type	indoline
ε /Mol^{-1}·cm^{-1}	λ_{max}/nm
43300 EtOH	at 483 EtOH
HOMO/V vs SCE	LUMO/V vs SCE
Structure	$C_{24}H_{22}N_2O_4S_2$ Exact Mass : 466.10 Mol. Wt. : 466.57 C, 61.78 ; H, 4.75 ; N, 6.00 ; O, 13.72 ; S, 13.75
ref.	Tamotsu Horiuchi, Hidetoshi Miura, Satoshi Uchida Highly efficient metal-free organic dyes for dye-sensitized solar cells J. Photochem. Photobio. A : Chem., 164, 29 (2004)
lab.	MPM & Uchida
remark	mp. 236℃ red/solution, red/on TiO$_2$

Name	D102
IUPAC Name	(5-{1,2,3,3a,4,8b-hexahydro-4-[4-(2,2-diphenylvinyl)phenyl]-cyclopeanta[b]indole-7-ylmethylene}-4-oxo-2-thioxo-thiazolidin-3-yl)acitic acid)
Type	indoline

ε /Mol^{-1}·cm^{-1}	λ_{max}/nm
55800 t-BuOH/AcN (1/1)	at 491
HOMO/V vs SCE	LUMO/V vs SCE

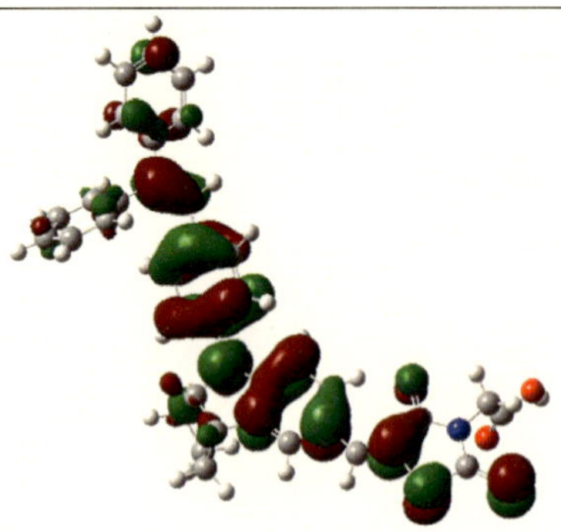

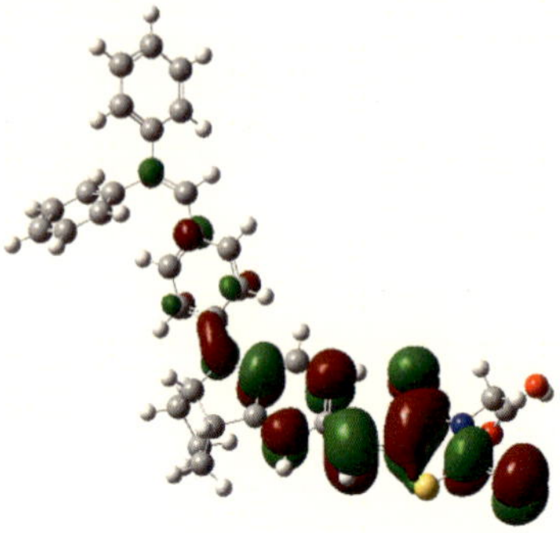

Structure

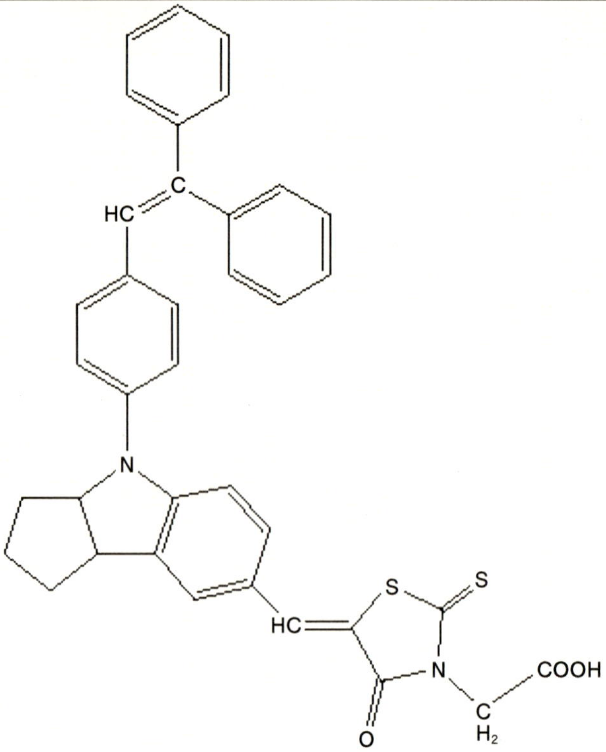

$C_{37}H_{30}N_2O_3S_2$
Exact Mass : 614.17
Mol. Wt. : 614.78
C, 72.29 ; H, 4.92 ; N, 4.56 ; O, 7.81 ; S, 10.43

ref.	Tamotsu Horiuchi, Hidetoshi Miura, Satoshi Uchida Highly-efficient metal-free organic dyes for dye-sensitized solar cells Chem. Comm., 3036 (2003) L. S-Mende et. al., Adv. Mater., 17, 813 (2005)
lab.	MPM & Uchida
remark	mp. >250℃ red/solution, red/on TiO_2

Name	D120
IUPAC Name	3-[1,2,3,3a,4,8b-hexahydro-4-(4-methoxyphenyl)-cyclopeanta[b]indole-7-yl]-2-cyano-acryl acid
Type	indoline
ε /Mol$^{-1}\cdot$cm^{-1}	λ_{max}/nm
40400t-BuOH/AcN (1/1)	at 390nm
HOMO/V vs SCE	LUMO/V vs SCE
Structure	(structure of D120) $C_{22}H_{20}N_2O_3$ Exact Mass : 360.15 Mol. Wt. : 360.41 C, 73.32 ; H, 5.59 ; N, 7.77 ; O, 13.32
ref.	Tamotsu Horiuchi, Hidetoshi Miura, Satoshi Uchida Highly efficient metal-free organic dyes for dye-sensitized solar cells Journal of Photochemistry and Photobiology A : Chemistry, 164, 29 (2004) Tamotsu Horiuchi, Hidetoshi Miura, Satoshi Uchida Highly efficient metal-free organic dyes for dye-sensitized solar cells JPPA : Chem., 164, 29 (2004)
lab.	MPM & Uchida
remark	red/solution, red/on TiO$_2$

Name	D131
IUPAC Name	3-{1,2,3,3a,4,8b-hexahydro-4-[4-(2,2-diphenylvinyl)phenyl]-cyclopeanta[b]indole-7-yl}-2-cyano-acryl acid
Type	indoline

ε /Mol$^{-1}\cdot$cm^{-1}	λ_{max}/nm
50000 MeOH	at 425 MeOH
HOMO/V vs SCE	LUMO/V vs SCE

Structure	C₃₅H₂₈N₂O₂ Exact Mass : 508.22 Mol. Wt. : 508.61 C, 82.65 ; H, 5.55 ; N, 5.51 ; O, 6.29
ref.	W. H. Howie, F. Claeyssens, H. Miura, L. M. Peter Characterization of Solid-State Dye-Sensitized Solar Cells Utilizing High Absorption Coefficient Metal-Free Organic Dyes J. Am. Chem. Soc., 130, 1367 (2008)
lab.	MPM & Uchida
remark	mp. >226℃ yellow/solution, yellow/on TiO$_2$

Name	D149
IUPAC Name	(5-{1,2,3,3a,4,8b-hexahydro-4-[4-(2,2-diphenylvinyl)phenyl]-cyclopeanta[b]indole-7-ylmethylene}-3'-ethyl-4,4'-dioxo-2'-thioxo-[2,5']bithiazolidine-3-yl)acetic acid
Type	indoline

ε /Mol^{-1}·cm^{-1}	λ_{max}/nm
68700 t-BuOH/AcN (1/1)	at 526 (at 541 on TiO$_2$)
HOMO/V vs SCE	LUMO/V vs SCE

Structure	$C_{42}H_{35}N_3O_4S_3$ Exact Mass : 741.18 Mol. Wt. : 741.94 C, 67.99 ; H, 4.75 ; N, 5.66 ; O, 8.63 ; S, 12.97
ref.	Tamotsu Horiuchi, Hidetoshi Miura, Kouichi Sumioka, Satoshi Uchida High Efficiency of Dye-Sensitized Solar Cells Based on Metal-Free Indoline Dyes J. Am. Chem. Soc., 126, 12218 (2004) S. Ito, S. M. Zakeeruddin, R. Humphry-Baker, P. Liska, R. Charvet, P. Comte, M. K. Nazeeruddin, P. Péchy, M. Takata, H. Miura, S. Uchida, and M. Grätzel, High-Efficiency Organic-Dye-Sensitized Solar Cells Controlled by Nanocrystalline-TiO$_2$ Electrode Thickness, *Adv. Mater.*, **18**, 1202 (2006)
lab.	MPM & Uchida
remark	wine red/solution, wine red/on TiO$_2$

Name	D150
IUPAC Name	(3"-carboxymethyl-5-{1,2,3,3a,4,8b-hexahydro-4-[4-(2,2-diphenylvinyl)phenyl]-cyclopeanta[b]indole-7-ylmeteylene}-3"-eteyl-4,4',4"-trioxo-2"-thioxo-[2,5 ; 2',5"]terthiazolidine-3'-yl-)acetic acid
Type	indoline
ε /Mol$^{-1}\cdot$cm^{-1}	λ_{max}/nm
HOMO/V vs SCE	LUMO/V vs SCE
Structure	C$_{47}$H$_{40}$N$_4$O$_5$S$_4$ Exact Mass : 868.19 Mol. Wt. : 869.11 C, 64.95 ; H, 4.64 ; N, 6.45 ; O, 9.20 ; S, 14.76
ref.	Tamotsu Horiuchi, Hidetoshi Miura, Kouichi Sumioka, Satoshi Uchida High Efficiency of Dye-Sensitized Solar Cells Based on Metal-Free Indoline Dyes J. Am. Chem. Soc., 126, 12218 (2004)
lab.	MPM & Uchida
remark	dark blue purple/solution, black/on TiO$_2$

色素増感太陽電池研究者のための色素データ集

Name	HRS-1
IUPAC Name	*cis*-di(thiocyanato)-(2,2'-bipyridyl-4,4'-dicarboxylic acid)(4,4'-di(hexylthienylvinyl)-2,2'-bipyridyl)-ruthenium(Ⅱ)
Type	Ru complex
ε /Mol^{-1}·cm^{-1}	λ_{max}/nm
42400 EtOH 18700 EtOH	at 372 at 542
HOMO/V vs SCE	LUMO/V vs SCE
Structure	$C_{48}H_{48}N_6O_4RuS_4$ Exact Mass：1002.17 Mol. Wt.：1002.27 C, 57.52；H, 4.83；N, 8.39；O, 6.39；Ru, 10.08；S, 12.80
ref.	Ke-Jian Jiang, Naruhiko Masaki, Jiang-bin Xia, Shuji Noda, Shozo Yanagida A novel ruthenium sensitizer with a hydrophobic 2-thiophen-2-yl-vinyl-conjugated bipyridyl ligand for effective dye sensitized TiO$_2$ solar cells Chem. Comm., 2460 (2006)
lab.	Yanagida
remark	

Name	JK-1
IUPAC Name	3-{5-[N,N-bis(9,9-dimethylfluorene-2-yl)phenyl]-thiophene-2-yl}-2-cyano-acrylic acid
Type	organic

ε /Mol^{-1}·cm^{-1}	λ_{max}/nm
34000 EtOH	at 354
30000 EtOH	at 436
HOMO/V vs SCE	LUMO/V vs SCE
0.92 on TiO$_2$	−1.58 on TiO$_2$

Structure	$C_{44}H_{34}N_2O_2S$ Exact Mass : 654.23 Mol. Wt. : 654.82 C, 80.70 ; H, 5.23 ; N, 4.28 ; O, 4.89 ; S, 4.90
ref.	Sanghoon Kim, Kee-hyung Song, Sang Ook Kang, Jaejung Ko The role of borole in a fully conjugated electron-rich system Chem. Comm., 68 (2004) Sanghoon Kim, Jae Kwan Lee, Sang Ook Kang, Jaejung Ko, J.-H. Yum, Simona Fantacci, Filippo De Angelis, D. Di Censo, Md. K. Nazeeruddin, Michael Grätzel Molecular Engineering of Organic Sensitizers for Solar Cell Applications J. Am. Chem. Soc., 128, 16701 (2006)
lab.	Ko
remark	

色素増感太陽電池研究者のための色素データ集

Name	JK-2
IUPAC Name	3-{5'-[N,N-bis-(9,9-dimethylfluorene-2-yl)phenyl]-2,2'-bisthiophene-5-yl}-2-cyano-acrylic acid
Type	organic
$\varepsilon / \mathrm{Mol}^{-1} \cdot \mathrm{cm}^{-1}$	$\lambda_{max}/\mathrm{nm}$
44000 EtOH	at 364
39000 EtOH	at 452
HOMO/V vs SCE	LUMO/V vs SCE
0.80 on TiO_2	-1.54 on TiO_2
Structure	$C_{48}H_{36}N_2O_2S_2$ Exact Mass : 736.22 Mol. Wt. : 736.94 C, 78.23 ; H, 4.92 ; N, 3.80 ; O, 4.34 ; S, 8.70
ref.	Sanghoon Kim, Kee-hyung Song, Sang Ook Kang, Jaejung Ko The role of borole in a fully conjugated electron-rich system Chem. Comm., 68 (2004) Sanghoon Kim, Jae Kwan Lee, Sang Ook Kang, Jaejung Ko, J.-H. Yum, Simona Fantacci, Filippo De Angelis, D. Di Censo, Md. K. Nazeeruddin, Michael Grätzel Molecular Engineering of Organic Sensitizers for Solar Cell Applications J. Am. Chem. Soc., 128, 16701 (2006)
lab.	Ko
remark	

Name	D190
IUPAC Name	(5-{1,2,3,3a,4,8b-hexahydro-4-[4-(4,4-diphenylbutadienyl)phenyl]-cyclopeanta[b]indole-7-ylmethylene}-3'-ethyl-4,4'-dioxo-2'-thioxo-[2,5']bithiazolidine-3-yl)acetic acid
Type	indoline
ε /Mol^{-1}·cm^{-1}	λ_{max}/nm
HOMO/V vs SCE	LUMO/V vs SCE
Structure	$C_{44}H_{37}N_3O_4S_3$ Exact Mass : 767.19 Mol. Wt. : 767.98 C, 68.81 ; H, 4.86 ; N, 5.47 ; O, 8.33 ; S, 12.53
ref.	2005年 電気化学秋季大会 1E03 「インドリン骨格を有する色素により増感された全固体色素増感型光電変換素子」 髙岡和千代, 石井康憲, 髙田昌和, 内田聡
lab.	MPM & Uchida
remark	red/solution, purple/on TiO$_2$

色素増感太陽電池研究者のための色素データ集

Name	D205
IUPAC Name	(5-{1,2,3,3a,4,8b-hexahydro-4-[4-(2,2-diphenylvinyl)phenyl]-cyclopeanta[b]indole-7-ylmethylene}-3'-n-octyl-4,4'-dioxo-2'-thioxo-[2,5']bithiazolidine-3-yl-)acetic acid
Type	indoline

ε /Mol$^{-1}\cdot$cm^{-1}	λ_{max}/nm
53000 THF	At 532 THF
HOMO/V vs SCE	LUMO/V vs SCE

Structure	$C_{48}H_{47}N_3O_4S_3$ Exact Mass：825.27 Mol. Wt.：826.10 C, 69.79 ; H, 5.73 ; N, 5.09 ; O, 7.75 ; S, 11.64
ref.	D. Kuang, S. Uchida, R. Humphry-Baker, S. M. Zakeeruddin and M. Grätzel, Organic Dye-Sensitized Ionic Liquid Based Solar Cells：Remarkable Enhancement in Performance through Molecular Design of Indoline Sensitizers, *Angew. Chem. Int. Ed.,* **47**, 1923 (2008)
lab.	MPM & Uchida
remark	red/solution, purple/on TiO$_2$

Name	ZnTPP
IUPAC Name	Zinc 5,10,15,20-tetrakis(4-carboxyphenyl)-21H,23H-porphine
Type	
ε /Mol^{-1}·cm^{-1}	λ_{max}/nm
HOMO/V vs SCE	LUMO/V vs SCE
Structure	C$_{56}$H$_{52}$N$_4$O$_8$Zn Exact Mass : 972.31 Mol. Wt. : 974.44 C, 69.02 ; H, 5.38 ; N, 5.75 ; O, 13.14 ; Zn, 6.71
ref.	K. Kalyanasundaram, N. Vlachopoulos, V. Krishnan, A. Monnier,t and M. Grätzel, Sensitlzation of TiO, in the Visible Light Region Using Zinc Porphyrins, *J. Phys. Chem.*, **91**, 2342 (1987)
lab.	Grätzel
remark	

Name	H$_2$TC$_1$PP
IUPAC Name	5,10,15,20-tetrakis(4-carboxyphenyl)-21H,23H-porphine
Type	
ε /Mol$^{-1}\cdot$cm^{-1}	λ_{max}/nm
HOMO/V vs SCE	LUMO/V vs SCE
Structure	C$_{48}$H$_{30}$N$_4$O$_8$ Exact Mass：790.21 Mol. Wt.：790.77 C, 72.90；H, 3.82；N, 7.09；O, 16.19
ref.	A. Kay and M. Grätzel, Artificial photosynthesis. 1. Photosensitization of titania solar cells with chlorophyll derivatives and related natural porphyrins, *J. Phys. Chem. B*, **97**, 6272 (1993)
lab.	Grätzel
remark	

Name	H$_2$TC$_4$PP
IUPAC Name	5-(4-carboxyphenyl)-10,15,20-tri-*tert*-butyl-21*H*,23*H*-porphine
Type	

ε /Mol^{-1}·cm^{-1}		λ_{max}/nm	

HOMO/V vs SCE		LUMO/V vs SCE	

Structure	C$_{57}$H$_{54}$N$_4$O$_2$ Exact Mass : 826.42 Mol. Wt. : 827.06 C, 82.78 : H, 6.58 : N, 6.77 : O, 3.87
ref.	特開 2006-032260 号公報
lab.	AIST
remark	

色素増感太陽電池研究者のための色素データ集

Name	Phthalocyanine Dye
IUPAC Name	Zinc phthalocyanine-2,9,16,23-tetra-carboxylic acid
Type	
ε /Mol^{-1}·cm^{-1}	λ_{max}/nm
HOMO/V vs SCE	LUMO/V vs SCE
Structure	$C_{44}H_{40}N_8O_8Zn$ Exact Mass : 872.23 Mol. Wt. : 874.25 C, 60.45 ; H, 4.61 ; N, 12.82 ; O, 14.64 ; Zn, 7.48
ref.	Y-C. Shen, L. Wang, Z. Lu, Y. Wei, Q. Zhou, H. Mao and H. Xu, Fabrication, characterization and photovoltaic study of a TiO$_2$ microporous electrode, *Thin Solid Films* 257, 144 (1995)
lab.	中国東南大学　中国科学院
remark	

Name	Phthalocyanine Dye
IUPAC Name	2-[2'-(zinc9',16',23'-tri-tert-butyl-29H,31H-phthalocyanyl)]succinic acid
Type	

ε /Mol$^{-1}\cdot$cm^{-1}		λ_{max}/nm	

HOMO/V vs SCE		LUMO/V vs SCE	

Structure	$C_{56}H_{68}N_8O_4Zn$ Exact Mass: 980.47 Mol. Wt.: 982.60 C, 68.45; H, 6.98; N, 11.40; O, 6.51; Zn, 6.66
ref.	P. Y. Reddy, L. Giribabu, C. Lyness, H. J. Snaith, C. Vijaykumar, M. Chandrasekharam, M. Lakshmikantam, J-H Yum, K. Kalyanasundaram, M. Grätzel, M. K. Nazeeruddin, Efficient Sensitization of Nanocrystalline TiO$_2$ Films by a Near-IR-Absorbing Unsymmetrical Zinc Phthalocyanine, *Angew. Chem. Int. Ed.*, **46**, 373 (2007)
lab.	中国東南大学　中国科学院
remark	

色素増感太陽電池研究者のための色素データ集

Name	Phthalocyanine Dye (TT-1)
IUPAC Name	Zinc phthalocyanine-tri-tert-butyl-2-carboxylic acid
Type	

ε /Mol^{-1}·cm^{-1}	λ_{max}/nm

HOMO/V vs SCE	LUMO/V vs SCE

Structure	$C_{56}H_{76}N_8O_2Zn$ Exact Mass : 956.54 Mol. Wt. : 958.67 C, 70.16 ; H, 7.99 ; N, 11.69 ; O, 3.34 ; Zn, 6.82
ref.	J-H. Yum, S-r. Jang, R. Humphry-Baker, M. Grätzel, J-J. Cid, T. Torres and Md. K. Nazeeruddin, Effect of Coadsorbent on the Photovoltaic Performance of Zinc Pthalocyanine-Sensitized Solar Cells, *Langumuir*, **24**, 5436 (2008)
lab.	UniVersidad Autónoma de Madrid & Grätzel
remark	

Name	Pendant type Polymer		
IUPAC Name			
Type			
ε /Mol^{-1}·cm^{-1}		λ_{max}/nm	
HOMO/V vs SCE		LUMO/V vs SCE	
Structure			
ref.	H. Osora, W. Li, L. Otero and M. A. Fox, Photosensitization of nanocrystalline TiO$_2$ thin films by a polyimide bearing pendent substituted-Ru (bpy)$_3^{+2}$ groups, *J. Photochem. Photobiol. B*, **43**, 232 (1998)		
lab.	University of Texas		
remark			

色素増感太陽電池研究者のための色素データ集

Name	Polythiophene Dye (P3TTA)		
IUPAC Name	poly-3-thiophenylacetic acid		
Type			
ε /Mol^{-1}·cm^{-1}		λ_{max}/nm	
HOMO/V vs SCE		LUMO/V vs SCE	
Structure	\multicolumn{3}{l}{ [structure of poly-3-thiophenylacetic acid with COOH group] }
ref.	G. K. R. Senadeera, K. Nakamura, T. Kitamura, Y. Wada and S. Yanagida, Fabrication of highly efficient polythiophene-sensitized metal oxide photovoltaic cells, *Appl. Phys. Lett.*, **83**, 5470 (2003)		
lab.	Yanagida		
remark			

Name	Cyanine Dye (C1-D)	
IUPAC Name		
Type		
ε /Mol^{-1}·cm^{-1}		λ_{max}/nm
HOMO/V vs SCE		LUMO/V vs SCE
Structure	\{structure image\} $C_{27}H_{29}N_2O_4$ Exact Mass : 445.21 Mol. Wt. : 445.53 C, 72.79 ; H, 6.56 ; N, 6.29 ; O, 14.36	
ref.	K. Sayama, K. Hara, Y. Ohga, A. Shinpou, S. Suga and H. Arakawa, Significant efects of the distance between the cyanine dye skeleton and the semiconductor surface on the photo-electrochemical properties of dye-sensitized porous semiconductor electrodes, *New. J. Chem.*, **25**, 200 (2001)	
lab.	Hayashibara & AIST	
remark		

色素増感太陽電池研究者のための色素データ集

Name	Cyanine Dye (SQ-3)	
IUPAC Name		
Type		
ε /Mol$^{-1}\cdot$cm^{-1}		λ_{max}/nm
HOMO/V vs SCE		LUMO/V vs SCE
Structure	![structure] $C_{42}H_{46}N_4O_8S_2$ Exact Mass : 798.28 Mol. Wt. : 798.97 C, 63.14 ; H, 5.80 ; N, 7.01 ; O, 16.02 ; S, 8.03	
ref.	W. Zhao, Y. J. Hou, X. S. Wang, B. W. Zhang, Y. Cao, R. Yang, W. B. Wang and X. R. Xiao, Study on squarylium cyanine dyes for photoelectric conversion, *Sol. Ener. Mater. Sol. Cells*, **58**, 173 (1999).	
lab.	中国科学院	
remark		

Name	Cyanine Dye (B1)
IUPAC Name	
Type	

ε /Mol^{-1}·cm^{-1}		λ_{max}/nm	
HOMO/V vs SCE		LUMO/V vs SCE	

Structure	$C_{52}H_{66}N_4O_6$ Exact Mass : 842.50 Mol. Wt. : 843.10 C, 74.08 ; H, 7.89 ; N, 6.65 ; O, 11.39
ref.	A. Burke, L. Schmidt-Mende, S. Ito and M. Grätzel, A novel blue dye for near-IR dye-sensitised solar cell applications. *Chem. Commun.*, 234 (2007)
lab.	Grätzel
remark	

色素増感太陽電池研究者のための色素データ集

2008年7月31日　第1刷発行
2008年9月30日　第2刷発行

　　著　者　　堀内　保
　　　　　　　藤沢潤一
　　　　　　　内田　聡　　　　　　　　　（S0761）
　　発行者　　辻　賢司
　　発行所　　株式会社シーエムシー出版
　　　　　　　東京都千代田区内神田1-13-1（豊島屋ビル）
　　　　　　　電話 03(3293)2061
　　　　　　　大阪市中央区南新町1-2-4（椿本ビル）
　　　　　　　電話 06(4794)8234
　　　　　　　http://www.cmcbooks.co.jp/

〔印刷　倉敷印刷株式会社〕©T. Horiuchi, J. Fujisawa, S. Uchida, 2008
定価はカバーに表示してあります。
落丁・乱丁本はお取替えいたします。

本書の内容の一部あるいは全部を無断で複写（コピー）することは，
法律で認められた場合を除き，著作者および出版社の権利の侵害
になります。

ISBN978-4-7813-0053-5　C3054　¥30000E